41

Advances in Biochemical Engineering/ Biotechnology

Managing Editor: A. Fiechter

Microbial Bioproducts

With contributions by
G. P. Agarwal, L. de Boer, H. Brandl,
L. Dijkhuizen, R. C. Fuller, R. A. Gross,
H. Höke, A. Läufer, R.W. Lenz,
R. Müller, Ch. Syldatk

With 23 Figures and 34 Tables

Springer-Verlag
Berlin Heidelberg New York London
Paris Tokyo Hong Kong Barcelona

ISBN 3-540-52569-6 Springer-Verlag Berlin Heidelberg New York
ISBN 0-387-52569-6 Springer-Verlag New York Berlin Heidelberg

Library of Congress Catalog Coard Number 72-152360
Printed of Germany

Bookbinding: Lüderitz & Bauer, Berlin
2152/3020-543210

Table of Contents

Microbial and Enzymatic Processes for L-Phenylalanine Production
L. de Boer, L. Dijkhuizen 1

Production of Optically Pure D- and L-α-Amino Acids by Bioconversion of D,L-5-Monosubstituted Hydantoin Derivatives
Ch. Syldatk, A. Läufer, R. Müller, and H. Höke 29

Plastics from Bacteria and for Bacteria: Poly(β-Hydroxyalkanoates) as Natural, Biocompatible, and Biodegradable Polyesters
H. Brandl, R. A. Gross, R. W. Lenz, and R. C. Fuller . . . 77

Glycerol
G. P. Agarwal 95

Author Index Volumes 1–41 129

Microbial and Enzymatic Processes for L-Phenylalanine Production

L. de Boer and L. Dijkhuizen*
Department of Microbiology, University of Groningen, Kerklaan 30, 9751 NN Haren, The Netherlands

1 Introduction ... 1
2 Microbial Production of L-Phenylalanine ... 2
  2.1 Regulation of L-Phenylalanine Biosynthesis ... 3
  2.2 Strain Construction ... 5
    2.2.1 Aromatic Amino Acid Pathway ... 5
    2.2.2 Intermediary Metabolism ... 8
    2.2.3 Application of Recombinant DNA Technology ... 9
3 Degradation of L-Phenylalanine ... 9
4 Enzymatic Production of L-Phenylalanine ... 10
  4.1 L-Phenylalanine Ammonia-Lyase (PAL) ... 11
  4.2 L-Phenylalanine Dehydrogenase (PheDH) ... 14
  4.3 Aromatic Amino Acid Aminotransferase (AAT) ... 17
5 Conclusions ... 20
6 Acknowledgements ... 21
7 Abbreviations ... 21
8 References ... 22

The aromatic amino acid L-phenylalanine is one of the building blocks for the dipeptide sweetener α-aspartame. Considerable progress has been made in recent years towards the development of microbial and enzymatic processes for L-phenylalanine production. Biosynthesis of this aromatic amino acid occurs via a complex pathway and is carefully controlled. Therefore, an extensive program for strain construction generally is required in order to achieve high levels of L-phenylalanine overproduction. In addition, microorganisms may possess a variety of enzymes capable of degrading L-phenylalanine. The possible application of a number of these enzymes for the conversion of suitable precursors into L-phenylalanine has received considerable attention. This paper attempts to review the information currently available on both approaches.

1 Introduction

L-Phenylalanine is produced commercially by way of the chemical synthesis from benzaldehyde, glycine and acetic acid anhydride, through purification from protein hydrolysates, or via microbial and enzymatic processes. L-Phenylalanine is one of

* To whom correspondence should be addressed

Advances in Biochemical Engineering/
Biotechnology, Vol. 41
Managing Editor: A. Fiechter

the building blocks of the dipeptide sweetener aspartame, L-α-aspartyl-L-phenylalanine methylester [1, 2] and the market of aspartame, and also that of L-phenylalanine, is steadily growing. The projected demand for L-phenylalanine for aspartame production is presented in Fig. 1 [3]. The amino acid is also used for pharmaceutical purposes and as a food additive. The rapidly increasing demand for L-phenylalaninc and the advantage of stereospecific biosynthesis have stimulated investigations into the possible production of this amino acid via microbial as well as enzymatic processes.

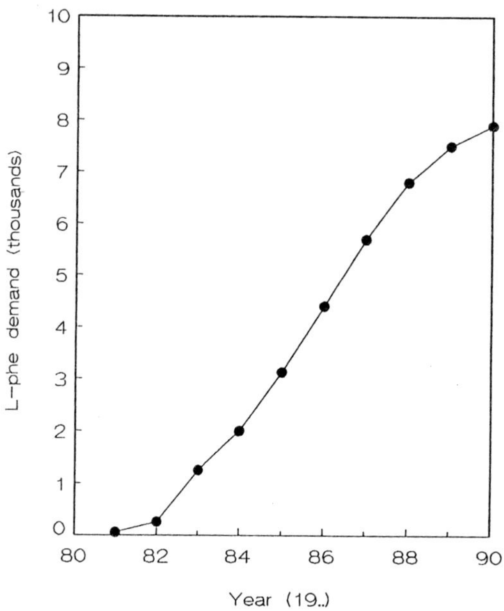

Fig. 1. Demand for L-phenylalanine (expressed in metric tons) for aspartame production [3]

2 Microbial Production of L-Phenylalanine

Synthesis of L-phenylalanine is energetically expensive [4]. Not surprisingly, this process is controlled accurately to meet the cellular demand and no phenylalanine overproducing bacterial strains have been isolated from the natural environment. The selection of a suitable putative production organism thus has to be based on alternative criteria. Conceivably, these should involve: ability of the organism to grow rapidly in mineral media without any requirement for expensive vitamins or other supplements; non-pathogenicity and absence of toxic products; inability to degrade L-phenylalanine and its precursors; sensitivity to inhibition of growth by phenylalanine analogs; availability of methods for the isolation of stable mutants and for further genetic manipulations required during strain development. The choice of the organism may also depend to some extent on its ability to use specific substrates as carbon- and energy sources for growth

[5–9]. The abundant availability, low cost and high levels of purity of methanol [8] makes this compound for instance an attractive feedstock for bioprocesses. Moreover, it may be argued that those methylotrophic bacteria employing the ribulose monophosphate (RuMP) pathway of formaldehyde assimilation offer the distinct additional advantage of possessing an unique metabolic pathway leading to erythrose-4-phosphate (E4P) and phosphoenolpyruvate (PEP) which are precursors for the biosynthesis of the aromatic amino acids [6, 10, 11].

Microbial production processes for L-phenylalanine have been developed in recent years. In practice, the choice of organism often has been made on the basis of experience previously obtained in developing industrial processes for the production of other amino acids. Not surprisingly, most attention thus far has been focused on *Escherichia coli*, *Bacillus subtilis*, and various coryneform bacteria.

2.1 Regulation of L-Phenylalanine Biosynthesis

The biosynthesis of the aromatic amino acids phenylalanine, tryptophan, and tyrosine, and the control mechanisms involved in various microorganisms have been reviewed extensively [12–19]. The overall regulation of the multi-enzyme,

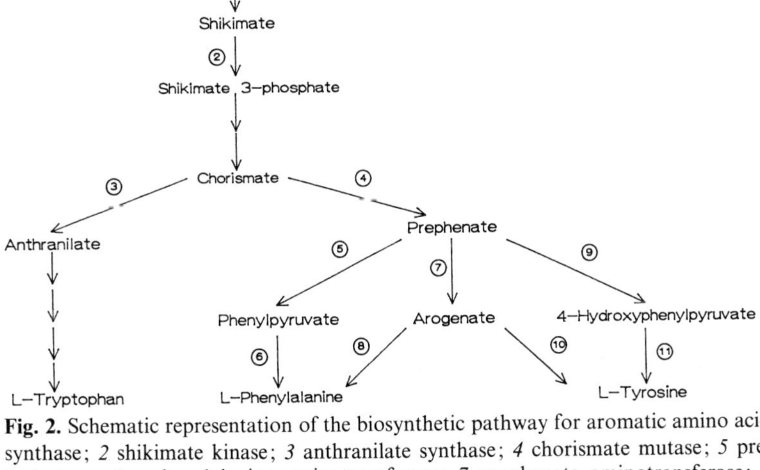

Fig. 2. Schematic representation of the biosynthetic pathway for aromatic amino acids. *1* DAHP synthase; *2* shikimate kinase; *3* anthranilate synthase; *4* chorismate mutase; *5* prephenate dehydratase; *6* L-phenylalanine aminotransferase; *7* prephenate aminotransferase; *8* arogenate dehydratase; *9* prephenate dehydrogenase; *10* arogenate dehydrogenase; *11* L-tyrosine aminotransferase

branched pathway (Fig. 2) is complex and may involve several isoenzyme systems, enzyme complexes, feedback regulation of key regulatory enzymes, both at the level of their synthesis (repression and attenuation) and activity (inhibition). Clearly, one of the main challenges in the development of bioprocesses for aromatic amino acids is the elucidation and subsequent effective deletion of the regulatory mechanisms involved in the organism under investigation.

The biosynthesis of all three aromatic amino acids starts with the formation of chorismate via shikimate in the shikimate pathway [20]. The pathway may also function in the synthesis of precursors for a variety of other aromatic compounds. The first step is the condensation of E4P and PEP, to yield 3-deoxy-D-arabino-heptulosonate 7-phosphate (DAHP), which is catalyzed by DAHP syn-

Table 1. Properties of regulatory enzymes involved in L-phenylalanine biosynthesis in various bacteria

Microorganism	Regulatory enzyme	Repressor	Inhibitor	$I_{0.5}$ (μM)	Inhibitor constant (μM)
Nocardia	DS	none	phe, tyr, trp	60, 60, 2	160, 180, 3
sp. 239	CM	none	phe, tyr	60, 30	60, 35
[11]	PDT	none	phe, tyr, trp	5, 15[c], 300	10, 20[a], 600
Escherichia	DS	phe, tyr,	phe, tyr	13, 82	13, 82
coli [33, 34–37]		trp[1]	trp[1]	none[1]	none[1]
	SK	tyr, trp	none	none	none
	CM	phe, tyr[1]	phe	50	—
	PDT	phe	phe	100	150
Bacillus	DS	tyr	chor, preph	200, 25	400, 50
subtilis	SK	tyr	chor, preph	—	—
[27, 38–43]	CM	phe, tyr, trp	preph	250	250
	PDT		phe, trp	30, 30	30, 30
			met, leu	3[c], 5[c2]	3[a], 2[a2]
Corynebacterium	DS	—	phe + tyr	—	1000[3]
glutamicum	CM	phe	phe, trp	50, 5[c]	60, 9[a]
[44–47]	PDT	—	phe, tyr, trp	2, <10[c], 25	2, <10[a], 25
Brevibacterium	DS	tyr	phe + tyr	51	—
flavum	CM	tyr	phe + tyr	50–750[b]	—
[48–52]	PDT	none	phe, tyr	2.5, 1.6[c]	1.0, 2.1[a]

DS, DAHP synthase; SK, shikimate kinase; CM, chorismate mutase; PDT, prephenate dehydratase; chor, chorismate; preph, prephenate; tyr, tyrosine; phe, phenylalanine; trp, tryptophan

[a] activator constant, defined as the activator concentration giving a two-fold increase in enzyme activity

[b] constant determined in the presence of 10 μm tryptophan

[c] $A_{0.5}$ (i.s.o. $I_{0.5}$) for an activator is defined as the activator concentration giving 50% of the maximum activation

[1] phe, tyr and trp feedback regulated isoenzymes, respectively

[2] expressed in mM

[3] at 37 °C

— no experimental data available

thase. Various isoenzymes of DAHP synthase may be present which are subject to feedback inhibition and repression by phenylalanine, tyrosine and tryptophan, and inhibition by intermediates of the pathway such as chorismate and phenylpyruvate [15, 16, 18, 21]. Only one DAHP synthase has been detected in the antibiotic producing actinomycetes that have been investigated. In these organisms DAHP synthase activity was only feedback inhibited to a minor extent, if at all, especially by L-tryptophan (22–26). In *E. coli* and *B. subtilis* feedback inhibition also occurred at the level of shikimate kinase [27–29].

Chorismate, the end product of the shikimate pathway, is converted into phenylalanine via the phenylpyruvate pathway, which is present for instance in *E. coli* and *B. subtilis*, or via dual pathways involving either phenylpyruvate or arogenate, as is the case in *Pseudomonas aeruginosa* [21, 30] (Fig. 2). In cyanobacteria, coryneform bacteria [14, 21] (e.g., *Corynebacterium* and *Brevibacterium* strains) and some sporeforming actinomycetes [31, 32], phenylalanine is synthesized exclusively via phenylpyruvate and tyrosine via arogenate. Especially chorismate mutase and prephenate dehydratase are targets for further control by phenylalanine and tyrosine [14, 16, 21]. In streptomycetes the carbon flow in the phenylalanine-specific branch is generally found to be feedback inhibited by phenylalanine at the level of prephenate dehydratase and tyrosine was shown to be an inhibitor or arogenate dehydrogenase in the tyrosine-specific branch [32].

The effectiveness and in vivo importance of the various control steps in aromatic amino acid biosynthesis can only be fully appreciated when considering the overall pattern of regulation of carbon flow over this branched biosynthetic pathway in a specific organism. This situation has only been achieved in a limited number of bacterial strains (Table 1). A comprehensive discussion of the regulation of a balanced synthesis of the three aromatic amino acids in *B. flavum* has been presented by Shiio [14]. The data provide a firm basis for the development of a suitable strategy for the construction of L-phenylalanine overproducing strains.

2.2 Strain Construction

Various examples of strain construction with the aim to develop microbial processes for phenylalanine overproduction are listed in Table 2. Feedback repression and/or inhibition control of enzymes in phenylalanine biosynthesis may be circumvented at least partially by using tyrosine auxotrophic mutants and supplying limiting amounts of this amino acid in the medium [53, 57, 58, 60, 62]. This offers the additional advantage of preventing undesirable accumulation of tyrosine from the common precursors. The remaining control mechanisms for phenylalanine biosynthesis may be eliminated by the stepwise isolation of mutants resistant to various amino acid analogs.

2.2.1 Aromatic Amino Acid Pathway

A methionine and tyrosine auxotrophic strain of *Brevibacterium lactofermentum* accumulated phenylalanine to a concentration of 4.2 g l^{-1} [62]. The successive

Table 2. Microbial production processes for L-phenylalanine

Microorganism	Main carbon source	L-phenylalanine production (g l^{-1})	Yield %	Cultivation time (h)	Ref.
Methylomonas methanolophila 6R E431 (β-2-TA^r 5-MT^r 3-AT^r)	Methanol	4	—	66	[8]
Bacillus polymyxa BT^r-7 (PFP^r β-2-TA^r)	Starch	0.5	5	72	[9]
Brevibacterium flavum 485-21 (MFP^r)	Glucose	2.2	2.2	72	[53]
Bacillus subtilis FF-25 (5-FT^r)	Glucose	6.0	7.5	48	[54]
Brevibacterium lactofermentum AJ3437 (PFP^r 5-MT^r Tyr$^-$, Met$^-$)	Glucose	22	17	72	[55]
Brevibacterium lactofermentum AJ11475 (PFP^r 5-MT^r DEC^s Tyr$^-$ Met$^-$)	Glucose	24.8	19	72	[56]
Brevibacterium flavum M-87 (PFP^r DS^r PDT^r Tyr$^-$, Met$^-$)	Glucose	23.4	18	72	[57]
Corynebacterium glutamicum 31-PAP-20-22 (PFP^r PAP^r Tyr$^-$)	Molasses	9.5	9.5	96	[58]
Corynebacterium glutamicum K38 (PFP^r MFP^r; genetically manipulated)	Molasses	19	19	100	[59]
Escherichia coli TA-6-7 Tyr134-7 (β-2-TA^r PFP^r Tyr$^-$)	Glucose	15	8.5	49	[60]
Corynebacterium sp. KY 7146 (Tyr$^-$)	n-Alkanes	10	15	68	[61]
Brevibacterium lactofermentum No. 123 (PFP^r β-3-TA^r adenine^r 5-MT^r Tyr$^-$ Met$^-$)	Glucose	21.7	16.6	72	[62]
Escherichia coli NST 74 (genetically manipulated)	Glucose	8.7	19	38	[63]

X^r, resistant to X; X$^-$, X$^-$ auxotroph; X^s, sensitive to X, MFP, M-fluorophenylalanine; PFP, P-fluorophenylalanine; 5-FT, 5-fluorotryptophan; 5-MT, 5-methyltryptophan; 3-AT, 3-aminotyrosine; PAP, P-aminotyrosine; β-2-TA, β-2-thienylalanine; β-3-TA, β-3-thienylalanine; DEC, decoyinine; DS^r, feedback inhibition resistant DAHP synthase; PDT^r, feedback inhibition resistant prephenate dehydratase introduction of resistance to *p*-fluoro-DL-phenylalanine (PEP), β-3-thienyl-D,L-alanine (β-3-TA), adenine and 5-methyl-DL-tryptophan (5-MT) resulted in the isolation of mutants producing 5.7, 8.8, 9.3, and 12.5 g l^{-1} phenylalanine, respectively. Optimization of the culture medium, especially by the addition of fumaric acid, allowed a further enhancement of phenylalanine production to 21.7 g l^{-1} after 72 h of cultivation. Using a L-phenylalanine overproducing mutant of *B. lactofermentum*, Akashi et al. demonstrated that maximum production of the amino acid occurred under conditions of a limited supply of oxygen [64]. The authors speculate that under these growth conditions less PEP is used for energy generation, resulting in an enhanced availability of this precursor for amino acid synthesis.

Following similar approaches, tyrosine auxotrophic and analog resistant mutants of *Corynebacterium glutamicum* were isolated producing 9.5 and 10 g l^{-1} of L-phenylalanine from cane molasses and n-alkanes (58, 61). The chorismate mutase and prephenate dehydratase genes (coding for feedback inhibition insensitive enzymes) of *C. glutamicum* K38, a PFP and *m*-fluoro-DL-phenylalanine (MFP) resistant strain have been cloned. Re-introduction of these genes on a recombinant plasmid into strain K38 via protoplast transformation further enhanced L-phenylalanine production from 13 to 19 g l^{-1} [59]. A PFP and β-2-TA resistant strain of *Bacillus polymyxa* was found to produce 0.5 g l^{-1} phenylalanine from starch [9]. Enzyme measurements revealed that this strain possessed a considerably increased level of prephenate dehydratase. This enzyme also had become insensitive to feedback inhibition by phenylalanine. Significant loss of the phenylalanine produced was observed towards the end of the production phase as the result of induction of the catabolic enzyme phenylalanine ammonia-lyase by phenylalanine.

In *Brevibacterium flavum* chorismate mutase and DAHP synthase form a bifunctional enzyme complex with common regulatory sites for phenylalanine and tyrosine. In MFP resistant *B. flavum* strains (possessing a feedback inhibition insensitive prephenate dehydratase), synergistic inhibition of DAHP synthase by MFP and tyrosine (formed intracellularly from the dipeptide Tyr-Glu in the medium) still occurred [53, 57]. The inhibitory effects on both DAHP synthase and chorismate mutase activities were simultaneously removed by inducing resistancy to MFP and Tyr-Glu, which resulted in an increase in phenylalanine production from 2.1 to 6.0 g l^{-1}. When following the same approach, phenylalanine production by strain No. 239, a PFP-resistant tyrosine/methionine double auxotrophic mutant, was enhanced from 18.1 to 23.4 g l^{-1} in strain M-87 [57].

Introduction of β-2-TA, 5-MT, and 3-aminotyrosine (3-AT) resistancies in *Methylomonas methanolophila* 6R, a methanol-utilizing bacterium [8], resulted in accumulation of all three aromatic amino acids, necessitating the isolation of tryptophan and tyrosine auxotrophic mutants in further work.

Overproduction of L-phenylalanine by strains of *E. coli* has been described by Choi and Tribe [63] and Park et al. [65]. A β-2-TA resistant strain of *E. coli* W3110 (strain TA-6–7) produced 5.7 g l^{-1} L-phenylalanine and a trace amount of tyrosine. Since tyrosine inhibited L-phenylalanine formation, a tyrosine auxotrophic mutant was isolated and this strain (TA-6–7 Tyr134) produced L-phenylalanine up to 11.4 g l^{-1}, with low tyrosine concentrations in the medium [65]. A PFP-resistant derivative strain was subsequently isolated and found to accumulate L-phenylalanine to a concentration of 15 g l^{-1} [60]. This strain was used to study optimal production conditions in a 500-liter pilot reactor [66].

Förberg and Häggström [67, 68] studied phenylalanine production by *E. coli* strains containing the recombinant plasmid pJN6, carrying genes for DAHP synthase (*aroF*) and feedback inhibition insensitive chorismate mutase/prephenate dehydratase (*pheA*). The effects of various regimes of glucose, tyrosine, sulphate and phosphate addition in the feed on phenylalanine production were investigated in (fed-)batch and continuous cultures. Exhaustion of phosphate in batch culture resulted in an immediate decrease in phenylalanine production, followed by a phase of slow product formation. The decrease of phenylalanine production was not so dramatic following depletion of sulphate. In the chemostat experiments phenylalanine production continued during phosphate limitation while sulphate and glucose limitation caused a collapse in the specific rate of product formation. It remains to be established whether these differences are purely due to physiological factors or to variations in plasmid stability with varying growth conditions.

In a similar approach, using a temperature-controllable expression vector carrying *aroF* and *pheA* genes encoding feedback inhibition insensitive enzymes, Sugimoto et al. [69] studied the temperature dependency of gene expression and phenylalanine production in *E. coli*. The concentration of phenylalanine was temperature dependent and highest production (18 g l^{-1}) was obtained at 38.5 °C in a 2.5–l reactor.

2.2.2 Intermediary Metabolism

A relatively novel aspect that draws increasing attention in studies with especially *B. flavum* and *C. glutamicum* is the possible enhancement of intracellular precursor concentrations to enlarge the carbon flow towards desired amino acids [70–77]. PEP is an important intermediate in central metabolic pathways. It is a precursor for the biosynthesis of various amino acids (lysine, aromatic amino acids) and intracellular PEP consumption other than for production of desired metabolites therefore should be minimized. Sugar metabolism in the amino acid producing bacterium *B. flavum* has been investigated in detail [70, 73]. Growth on several sugars involved a PEP dependent sugar phosphotransferase system (PTS). The presence of other PEP consuming enzymes, e.g., oxaloacetate decarboxylase, malic enzyme, and pyruvate kinase was investigated but only the latter enzyme, together with the PTS system, was detectable at growth supporting levels. A pyruvate kinase minus mutant was still able to grow under conditions where the PTS system was operative, thus generating the pyruvate required. Interestingly, mutational inactivation of pyruvate kinase in lysine [71] and aspartate [72] overproducing strains of *B. flavum* resulted in increased productivity. As outlined above, PEP is also a precursor for aromatic amino acid biosynthesis. Conceivably, similar approaches might also be applicable for further enhancing L-phenylalanine biosynthesis.

2.2.3 Application of Recombinant DNA Technology

The rate of amino acid biosynthesis may be increased considerably by the introduction of structural genes on multi-copy plasmids [5, 78–80]. At the moment there is limited information only in the literature about the application of recombinant DNA technology for the production of L-phenylalanine [59, 63, 67–69]. Detailed studies on the molecular biology and genetic regulation of aromatic amino acid synthesis thus far mainly have been restricted to *E. coli* (reviewed in Refs. [15, 16, 81–83]). The phenylalanine operon in *E. coli*, encoding chorismate mutase and prephenate dehydratase (*pheA*; [84]) and the genes encoding the three DAHP synthase isoenzymes (*aroF*, *aroG*, *aroH*; for a comparison, see Ref. [85]) have been cloned and sequenced. In view of the well-developed *E. coli* genetics it is not surprising that application of recombinant DNA technology for the construction of strains overproducing amino acids (L-phenylalanine [69] and several other amino acids [86, 87]) has been successful especially with this organism. Nevertheless, rapid progress also has been made in recent years towards the development of recombinant DNA technology for coryneform bacteria, including transformation, transfection methods, the development of suitable host-vector systems, and protoplast fusion techniques (for reviews, see Refs. [86, 88–91]). Moreover, genes coding for enzymes of the common aromatic amino acid biosynthetic pathway in *B. lactofermentum* [92], chorismate mutase [59] and prephenate dehydratase of *C. glutamicum* [59, 93] already have been cloned by complementation of auxotrophic mutants of the same organisms.

The analysis of the molecular basis for mutations in structural genes and regulatory sequences, causing desensitization of enzymes to feedback inhibition and increased levels of gene expression, respectively, is just beginning [85, 94, 95]. Conceivably, this is another area where recombinant DNA technology may significantly contribute to a further enhancement of amino acid production.

3 Degradation of L-Phenylalanine

Many bacteria are versatile organisms, amongst others able to grow on a variety of amino acids as carbon-, energy- and/or nitrogen source. Synthesis of catabolic enzymes during microbial amino acid production may result in a considerable decrease in yield and productivity. In many production strains synthesis of catabolic enzymes is inducible and repressed in the presence of alternative carbon sources such as glucose. Degradation of the amino acid to be overproduced may nevertheless still occur following depletion of alternative substrates towards the end of the process. It thus follows that the early identification and efficient deletion of catabolic enzymes via mutation clearly is of importance during strain construction.

Metabolic pathways for L-phenylalanine degradation are diverse. Whereas the last step in the biosynthesis of phenylalanine is generally catalyzed by aromatic amino acid aminotransferases [13], the initial step in the catabolic pathways may involve either phenylalanine hydroxylase [96–98], phenylalanine

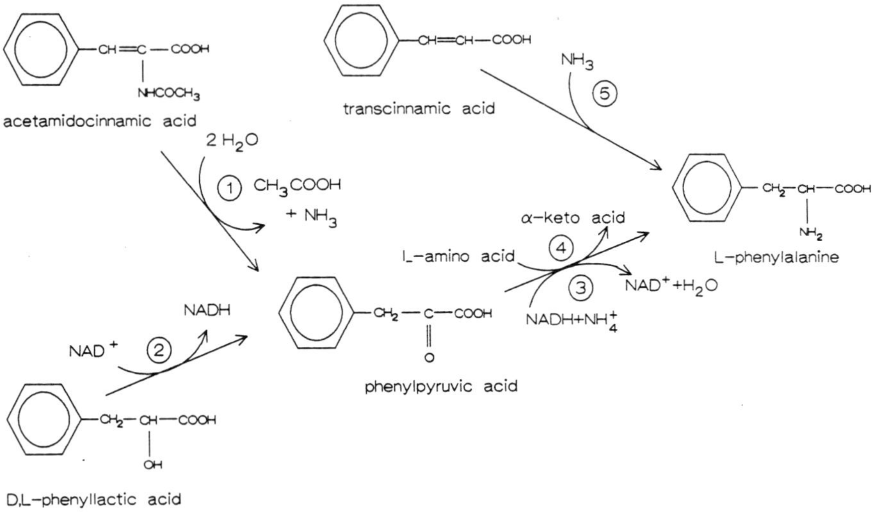

Fig. 3. Enzymatic reactions leading to L-phenylalanine production from phenylpyruvate, acetamidocinnamic acid and D,L-phenyllactic acid. *1* ACA acylase; *2* D- and L-hydroxyisocaproate dehydrogenase; *3* L-phenylalanine dehydrogenase; *4* aromatic amino acid aminotransferase; *5* phenylalanine ammonia-lyase

ammonia-lyase (PAL; Sect. 4.1), NAD-dependent phenylalanine dehydrogenase (PheDH; Sect. 4.2), or an aromatic amino acid aminotransferase (AAT; Sect. 4.3). The reactions catalyzed by some of these enzymes are shown in Fig. 3. It should be realized that phenylpyruvate, produced from phenylalanine by PheDH and AAT activities, also is an intermediate in phenylalanine biosynthesis (Fig. 2). The possible utilization of this compound as a growth substrate by the production strain therefore should be excluded as well. This will necessitate further mutational inactivation of phenylpyruvate catabolic enzymes, e.g., phenylpyruvate decarboxylase [99].

4 Enzymatic Production of L-Phenylalanine

Enzymatic synthesis of L-phenylalanine has been described starting out with racemic mixtures of D,L-phenylalanine (chemically produced) or by conversion of achiral precursors. Resolution of racemic mixtures may involve aminoacylases (from *N*-acetyl-DL-phenylalanine), hydantoinases (from D,L-hydantoins), esterases (from D,L-phenylalanine esters) or aminopeptidases (from D,L-phenylalanine amide). Various aspects of the stereospecific resolution of amino acids from racemic mixtures have been reviewed previously [100–102].

The conversion of chemically synthesized achiral precursors into L-phenylalanine, in single or in several enzymatic steps, has been studied in detail in the last three decades [103]. Several enzymatic processes, either with whole cells or purified enzymes, have been described which at least potentially may find an application (see below). In general, research focusses on the identification of cheap, novel substrates for biotransformations, the isolation of organisms which produce highest levels of the enzymes of choice, and most suitable reaction conditions. The enzymatic approach requires that process conditions are carefully controlled, in order to ensure prolonged stability of substrates and biocatalysts, and may necessitate cofactor regeneration. In the following the properties of phenylalanine ammonia-lyase, phenylalanine dehydrogenase and aromatic amino acid aminotransferase, enzymes which have been studied in most detail for phenylalanine production, are reviewed. A general reaction scheme is presented in Fig. 3.

4.1 L-Phenylalanine Ammonia-Lyase (PAL)

The first report concerning the discovery of PAL, which catalyzes the non-oxidative deamination of L-phenylalanine into ammonia and transcinnamic acid under physiological conditions, was published in 1961 [104]. Subsequent work demonstrated PAL activity in plants [104, 105], yeasts [106–110], fungi [111, 112], and bacteria [9, 113, 114]. Instead of pyridoxal 5′-phosphate, the cofactor in many amino acid transforming enzymes, PAL contains dihydroalanine as a prosthetic group. In plants PAL especially functions in secondary metabolism; it constitutes the first step of a highly branched pathway for the synthesis of lignins and related polyphenols. In various yeasts, fungi and bacteria the enzyme catalyzes the first step in phenylalanine catabolism, to generate carbon, energy and nitrogen

Table 3. Enzymatic production of L-phenylalanine from transcinnamic acid using phenylalanine ammonia-lyase as a biocatalyst

Microorganism	Cinnamic acid conc. (%)	L-Phenylalanine production (g l^{-1})	Maximum yield (%)	Processing time (h)	Ref.
Endomyces lindneri	4	32	71	48	[112]
Rhodotorula glutinis	2.2	18	70	25	[118]
Rhodotorula rubra FP10M6	1.75	17.8	91	37	[125]
Rhodotorula rubra SPA10	2	17.7	89	30	[126]
Rhodotorula rubra SPA10	1	9.4	84	24	[127]
Rhodotorula rubra SPA10	2	>17	88	30	[128]
Rhodotorula rubra FP10M6	4	50	83	120	[129]
Rhodotorula graminis	1.5	50.8	86	88	[130]
Rhodotorula rubra	5.5	59	90	—	[131]

for growth. The first suggested application for PAL was the removal of phenylalanine and tyrosine from blood and food to treat the symptoms of patients suffering of phenylketonuria [115]. For this purpose, PAL of the yeast *Rhodotorula rubra* was immobilized onto cellulose triacetate fibers. Extensive screening for PAL activity in yeasts has resulted in the isolation and identification of especially red pigmented strains of the genus *Rhodotorula* [116, 117]. L-Phenylalanine production from transcinnamic acid using PAL as a biocatalyst was first described in 1981 [118]. Growth and incubation conditions resulting in maximal PAL induction levels and rates of conversion of transcinnamic acid into phenylalanine by whole cells of *Rhodotorula glutinis* were examined [118, 119]. The equilibrium of the enzyme reaction strongly favours L-phenylalanine degradation under physiological conditions. Very high pH and ammonia concentrations, however, will drive the reaction in the direction of phenylalanine formation. Optimization of the reaction conditions resulted in accumulation of 18 g l^{-1} phenylalanine with a transcinnamic acid conversion yield of 70% (Table 3).

The main problems encountered with the application of PAL as a biocatalyst for L-phenylalanine production are its relatively low specific activity, instability and sensitivity to substrate inhibition. The regulation of PAL synthesis in yeasts has been studied in detail. Several amino acids were able to induce de novo synthesis of PAL in *R. glutinis* but maximum levels were obtained with L-phenylalanine [119]. Marusich et al. [120] observed that *R. glutinis* expressed a single PAL enzyme, regardless of whether the enzyme was induced during growth on L-phenylalanine as the carbon- or nitrogen source. Experiments on PFP resistant strains showed that PAL synthesis is controlled by a positive regulatory gene that affects expression of the entire phenylalanine catabolic pathway [121]. In *Neurospora crassa*, PAL activity was only detected under conditions of derepression for nitrogen, provided L-phenylalanine was present in the incubation medium [111]. Moreover, it was found that the product of a major regulatory locus, the *nit-2* gene, was involved in controlling expression of the enzyme. Studies with *Rhodosporidium toruloides* demonstrated that phenylalanine rather than an intermediate of the phenylalanine catabolic pathway acted as PAL inducer. Glucose was found to be a strong repressor and incubation of cells in a medium containing glucose, phenylalanine and ammonia even resulted in total repression of PAL synthesis [109]. Similar effects of glucose, ammonia and phenylalanine were observed in subsequent studies on the regulation of the synthesis of functional PAL mRNA [122]. Meanwhile, the gene encoding PAL in this organism has been cloned [123] and the complete nucleotide sequence determined [124]. Further optimization of PAL synthesis by *Rhodotorula* strains was reported by Evans et al. [117]. Highest PAL expression by *R. rubra* strain SPA10 was observed at temperatures and pH values which were suboptimal for growth. The presence of sugars and alcohols in the medium completely abolished PAL synthesis and also ammonia was found to repress the enzyme. Various mutants of *R. rubra* strain SPA10 resistant to phenylalanine analogs were isolated [125]. Attempts to isolate mutants producing PAL constitutively failed and only hyperactive PAL mutants, e.g., strain FP10M6 which displayed five times higher PAL activities compared to the parent strain, were obtained. Whole cells of this mutant produced 16.2 g l^{-1} L-phenylalanine from 17.5 g l^{-1} transcinnamic acid after 24 h of incubation.

The levels of the enzyme PAL in yeasts are further controlled by a mechanism of inactivation and significant loss of activity is consistently observed after peak production of the enzyme. The biocatalytic activity of the enzyme is prolonged by addition of L-isoleucine which probably acts by inhibiting the activity or synthesis of the PAL inactivating system [119]. The phenomenon of inactivation also is strongly delayed by addition of D,L-phenylalanine, by shifting the temperature to 10 °C and by making the incubation medium anaerobic [117, 126].

Phenylalanine production from transcinnamic acid has been studied in detail with *R. rubra* SPA10 (Table 3; [125–128]). Transcinnamic acid concentrations above 1% were inhibitory and conversion yields of 100, 69 and 47% were achieved with 1, 2, and 3% of transcinnamic acid, respectively. L-Glutamate was found to be a strong activator of PAL, whilst the presence of chloride ions greatly decreased the conversion efficiency. In the presence of $(NH_4)_2SO_4$ a 88% yield was obtained from 2% transcinnamate. The inhibitory effect of high transcinnamic acid concentrations was further diminished by the addition of glycerol and sorbitol [127]. The reuseability of PAL induced cells of *R. rubra* strains was generally low; the productivity rapidly decreased and varied strongly after an initial production period [125]. Various improvements subsequently were reported [126]. Addition of alginate, glutaraldehyde and polyethylene glycol (PEG) considerably enhanced stability of PAL during production. At pH 9.0, PAL, was considerably more stable than at the optimum pH for enzyme activity (pH 10.2). Oxygen appeared to stimulate the inactivation mechanism of PAL and sparging the incubation medium with nitrogen under otherwise optimal conditions significantly enhanced PAL stability.

L-Phenylalanine production by alginate immobilized whole cells of the hyperactive PAL mutant of *R. rubra*, strain FP10M6 (see above) was subsequently studied by Evans et al. [129]. When using conditions previously shown to stabilize PAL during enzyme induction (L-phenylalanine; L-isoleucine; transfer to anaerobic conditions at the end of the growth phase) and L-phenylalanine production (PEG, glutaraldehyde, polyhydric alcohols, and nitrogen sparging), a 80% conversion yield was obtained after 50 d of operation of a column bioreactor. This resulted in L-phenylalanine concentrations up to 50 g l^{-1} while retaining high reuseability of the biocatalyst.

Studies with *R. graminis* [130] revealed that this yeast possesses a considerably more stable PAL enzyme than *R. rubra*, both during growth and L-phenylalanine production from transcinnamic acid. Wild type *R. graminis*, however, displayed only relatively low activity levels of the enzyme. Among phenylpropiolic acid resistant mutants of *R. graminis*, strain GX6000 showed four- to five-fold higher PAL levels than the wild type. Induction of PAL synthesis in this mutant was less tightly regulated. L-Leucine acted synergistically with L-phenylalanine, resulting in 40% higher levels of PAL compared with the inducing effect of L-phenylalanine alone. In bioreactor experiments with this mutant, 50.8 g l^{-1} of L-phenylalanine was obtained in 88 h, with a transcinnamic acid conversion yield of

86%. A novel type of PAL enzyme, resistant to relatively high transcinnamic acid concentrations, also was detected in cells of *Endomyces lindneri* [112].

A large-scale commercial process for L-phenylalanine production from cinnamic acid with a PAL bioreactor has been reported by the Genex Corporation [131]. For this purpose a genetically manipulated strain of *R. rubra* that synthesized high levels of PAL was immobilized with vermiculate as the carrier and installed in a bioreactor. A solution containing 7.85 M ammonia, 0.37 M transcinnamate, at pH 10.3, was fed into the reactor and converted into L-phenylalanine (59 g l^{-1}) with a conversion yield of 90% (Table 3).

4.2 L-Phenylalanine Dehydrogenase (PheDH)

The first report on PheDH was published in 1984 [132] but the existence of this enzyme had already been suggested in 1965 [133]. The enzyme stereospecifically catalyzes the reductive amination (oxidative deamination) of phenylpyruvate (L-phenylalanine). PheDH activity was demonstrated first in an bacterial isolate, identified as a *Brevibacterium* species, using phenylalanine as the sole source of carbon and energy [132]. The presence of this enzyme sub-

Table 4. Characteristic properties of L-phenylalanine dehydrogenase from various microorganisms

Microorganism	Substrate specificity[a]	Molecular weight kDa	Number of subunits[b]	Optima[c]	
				pH	temp (°C)
Rhodococcus sp. M4 [134, 146]	PHP, KMB	—	—	9.5	50
Sporosarcina ureae [136, 137]	PHP, KMB, KC, KIC	305	8	9	40
Bacillus sphaericus [137]	PHP, KMB, KIV, KIC, KV	340	8 or 9	10.3	50
Bacillus badius [138]	PHP, KMP, KMB, KH, KIV	360	8	9.4	—
Rhodococcus maris K-18 [141]	PHP, KC, KMB, IP	70	2	9.8	—
Nocardia sp. 239 [143]	PHP, IP, KIC	42	1	10	53
Brevibacterium sp. [144, 145]	PHP, IP, KMB	—	—	9.3	52

[a] only substrates displaying at least 5% of the activity observed with phenylpyruvate are shown
[b] identical subunits in each case
[c] optima estimated for the reductive amination of phenylpyruvate
— = not determined
PHP, *p*-hydroxyphenylpyruvate; IP, indole-3-pyruvate; KMB, α-keto-γ-methylthiobutyrate; KMP, α-keto-γ-methylpentanoate; KH, α-ketohexanoate; KV, α-ketovalerate; KIV, α-ketoisovalerate; KC, α-ketocaproate; KIC, α-ketoisocaproate

sequently was also reported in various other Gram-positive bacteria, *Rhodococcus* sp. M4 [134], *Sporosarcina ureae* [135, 136], *Bacillus sphaericus* [137], *Bacillus badius* [138], *Corynebacterium equi* EVA-5 [139], *Micrococcus luteus* [140], *Rhodococcus maris* K-18 [141], *Thermoactinomyces intermedius* [142], and *Nocardia* sp. 239 [143]. Synthesis is induced by L-phenylalanine and the enzyme catalyzes the first step in L-phenylalanine catabolism during growth in media containing the amino acid either as a carbon- or nitrogen source. The enzyme from various bacteria has been purified and some characteristic properties are presented in Table 4. All enzymes exclusively use NAD as coenzyme and simple methods for the determination of L-phenylalanine and phenylpyruvate using PheDH have been described [142, 147]. The enzymes generally also posses considerable activity towards other substrates related to phenylpyruvate. The equilibrium of the PheDH reaction strongly favors L-phenylalanine formation. This enzyme has considerable commercial potential and worldwide investigations were started towards its application for phenylalanine production. Published data on L-phenylalanine production involving PheDH are presented in Table 5. The main disadvantages of this system are the requirement of extensive cofactor regeneration, biocatalyst instability and sensitivity to substrate inhibition by phenylpyruvate. Thermophilic bacteria constitute a promising source for biocatalysts of higher stability. The presence of thermostable PheDH in thermophilic actinomycetes has been reported recently [142, 151].

Table 5. Enzymatic production of L-phenylalanine from phenylpyruvate using phenylalanine dehydrogenase as a biocatalyst

Microorganism	Substrate	Biocatalyst	L-Phenylalanine production ($g\ l^{-1}$)	Maximum yield (%)
Rhodococcus sp. M4 [134]	PPA	Purified enzyme	456[a]	95
Sporosarcina ureae [136]	PPA	Purified enzyme	116	89
Corynebacterium equi OARI-16 [139]	ACA	Whole cells	32	99
Brevibacterium sp. [144]	PPA	Crude enzyme	37.4[a]	93
Rhodococcus sp. M4/ *Brevibacterium* sp. 37/3 [148]	ACA	Purified enzymes[b]	50.5	100
Lactobacillus confusus/*Lactobacillus casei*/*Rhodococcus* sp. M4 [149]	PLA	Purified enzymes[c]	28[a]	43
Nocardia opaca [150]	PPA, H_2	Whole cells	1.45	82

PPA, phenylpyruvic acid; ACA, acetamidocinnamic acid; PLA, DL-phenyllactic acid

[a] space-time yield ($g\ l^{-1}\ d^{-1}$)

[b] ACA acylase, PheDH, and formate dehydrogenase

[c] D- and L-hydroisocaproate dehydrogenases and PheDH

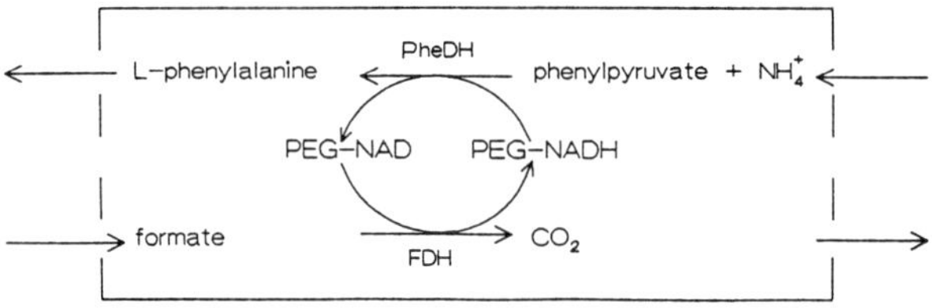

Fig. 4. Schematic presentation of an enzyme membrane reactor. FDH, formate dehydrogenase; PEG-NAD(H), polyethylene glycol linked NAD(H), PheDH, L-phenylalanine dehydrogenase; ------, ultrafiltration membrane

Following the isolation of *Brevibacterium* sp., growth conditions allowing highest PheDH production, and the kinetic parameters of the partially purified enzyme were studied [144]. Continuous L-phenylalanine production with coenzyme regeneration required construction of an enzyme-membrane reactor in which partially purified PheDH, formate dehydrogenase and PEG-linked NAD(H) were retained behind an ultrafiltration membrane (Fig. 4). Feeding a solution of phenylpyruvate, ammonia and formate into the reactor resulted in a space time yield of 37.4 g l^{-1} d^{-1} L-phenylalanine with an average conversion of 93%. A similar study was performed with *Rhodococcus* sp. M4 [134]. Compared to *Brevibacterium* sp. and *Sporosarcina ureae*, *Rhodococcus* sp. M4 clearly is a superior PheDH producer [152]. In addition, the enzyme of the latter organism was considerably more stable, and a space time yield of 456 g l^{-1} d^{-1} was achieved in an enzyme-membrane reactor. An on-line flow-injection monitoring system, to measure the inducing concentration of L-phenylalanine during continuous cultivation of *Rhodococcus* sp. M4, allowed further optimization of PheDH production [153]. Asano and Nakazawa [136] reported the production of the L-isomers of phenylalanine, tyrosine, tryptophan, methionine, valine, leucine, isoleucine, and *allo*-isoleucine from the respective α-keto acids, using purified PheDH of *Sporosarcina ureae* in a cofactor regenerating system based on formate dehydrogenase and catalytic amounts of NAD. Meanwhile, the gene coding for PheDH in *B. sphaericus* has been cloned and sequenced [154]. Introduction of the gene into *E. coli* resulted in a 40-fold higher total enzyme activity per liter of culture than in the *B. sphaericus* parent strain [155].

Conversion yields of phenylpyruvate into L-phenylalanine are negatively affected by instability of phenylpyruvate and, when provided at relatively high concentrations, its inhibitory effect on PheDH activity. This has stimulated the search for alternative, cheap precursors that easily can be converted into phenylpyruvate, followed by synthesis of L-phenylalanine via PheDH (or aromatic amino acid aminotransferase, see below) activity. Especially the potential application of racemic mixtures of D,L-phenyllactate and acetamidocinnamic acid (ACA) have been studied. Production of L-phenylalanine from ACA by immobilized whole cells of *C. equi* EVA-5 was reported by Evans et al. [139]. The data indicated that this process proceeds via α-aminocinnamate, α-iminocinnamate, α-aminophenyllactate, and phenylpyruvate as intermediates. The first step, conversion of ACA into α-aminocinnamic acid and acetate, was catalyzed by

ACA acylase while the reductive amination of phenylpyruvate into L-phenylalanine involved PheDH. L-Phenylalanine formation via PheDH requires the supply of reducing equivalents and the productivity of the system was considerably enhanced in the presence of oxidizable co-substrates such as lactate. Selection of PFP and OFP (*o*-fluoro-D,L-phenylalanine) resistant mutants resulted in the isolation of mutant OARI-16 which constitutively expressed high levels of ACA acylase and accumulated 32 g l^{-1} L-phenylalanine from 40 g l^{-1} ACA. Using purified ACA acylase from *Brevibacterium* strain 37/3, purified PheDH from *Rhodococcus* sp. M4 and commercial formate dehydrogenase, Hummel et al. [148] reported production of 50.5 g l^{-1} of L-phenylalanine within 75 h, with an ACA conversion yield close to 100%.

D,L-Phenyllactate can be converted into phenylpyruvate by the action of D- and L-hydroxyisocaproate dehydrogenases of *Lactobacillus casei* [156] and *Lactobacillus confusus* [157], respectively. Both phenyllactate converting enzymes were purified and immobilized behind an ultrafiltration membrane, together with PEG-NAD(H) and PheDH of *Rhodococcus* sp. M4. This resulted in the formation of L-phenylalanine with a space time yield of 28 g l^{-1} d^{-1} from 50 mM D,L-phenyllactate [149].

Matsunaga et al. [150] described production of L-phenylalanine from phenylpyruvate and ammonia with calcium alginate immobilized cells of *Nocardia opaca* incubated under high hydrogen pressure (100 atm). L-Phenylalanine synthesis involved both phenylalanine aminotransferase and PheDH activity.

4.3 Aromatic Amino Acid Aminotransferase (AAT)

The intracellular roles of bacterial AAT has been reviewed earlier [158]. Aminotransferases are involved in amino acid biosynthesis and catabolism, and secondary metabolism. These proteins contain pyridoxal 5′-phosphate as cofactor and generally display strict stereospecificity for either L- or D-isomers.

During growth of *Brevibacterium linens* [159–161], *Aspergillus niger* [162], *Pseudomonas putida* [163] and some cheese coryneform bacteria [164] on L-phenylalanine as carbon- and nitrogen source, inducible AAT enzymes were detected. The levels of these catabolic enzymes were much higher than during growth under conditions where similar AAT's merely acted in L-phenylalanine biosynthesis. Thermostable L-phenylalanine aminotransferases have been identified in various thermophilic bacteria [165]. The main disadvantages of application of AAT for L-phenylalanine production are the inhibition of enzyme activity by phenylpyruvate, the requirement for amino donors, and the equilibrium constant of the reaction (close to 1). This necessitates the continuous withdrawal of the α-keto acid produced from the amino donor.

The first report on AAT catalyzed production of L-phenylalanine from phenylpyruvate (Table 6) appeared in 1959 [103]. While screening for microorganisms capable of producing L-phenylalanine from phenylpyruvate and amino donors, highest activity was found with *Alcaligenes faecalis*. L-Phenylalanine was formed to a concentration of 6 g l^{-1} with a yield of 63.5%, using L-glutamate and

L-aspartate as amino donors. A 76% conversion of phenylpyruvate to phenylalanine was obtained with *Serratia marcescens* aminotransferase, using a glutamate regeneration system consisting of glutamate dehydrogenase, ethanol and alcohol dehydrogenase [166]. A disadvantage of this approach is that the acetaldehyde produced and the remaining ethanol have to be separated from the end-product. L-Phenylalanine production to 7.5 g l^{-1} from phenylpyruvate was obtained with a valine overproducing strain of *C. glutamicum*, containing high phenylalanine aminotransferase activity. Growth of this strain in a medium with

Table 6. Enzymatic production of L-phenylalanine from phenylpyruvate using aromatic amino acid (L-phenylalanine) aminotransferase as a biocatalyst

Microorganism	Substrate	Biocatalyst	L-Phenylalanine concentration (g l^{-1})	Maximum yield (%)
Alcaligenes faecalis [103]	PPA/asp/glu	Whole cells	6	63.5
Clostridium butyricum-alanine dehydrogenase/*Micrococcus luteus* [140]	PPA/H_2 pyruvate/NH_4Cl	Whole cells	12	80
Pseudomonas putida Z4 [163]	PPA/asp	Purified enzyme	7.9	79
Corynebacterium glutamicum [167]	PPA	Whole cells	7.5	75
Escherichia coli B Z1196 (*tyrB*) [168]	D-glucose/PPA	Resting cells	28.5	95
Pseudomonas fluorescens ATCC 11250 [170]	PPA/asp/glu	Immobilized cells	>17[b]	>85
Paracoccus denitrificans pFPr-1 [171]	PPA/asp	Immobilized cells	122[a] 22[b]	92.5 90
Escherichia coli B 11303 [172]	PPA/asp	Whole cells	30	98
Pseudomonas putida Z4 [173]	PPA/asp	Immobilized cells	6.6	65
		Purified enzyme	9.4	94
Bacillus sphaericus N-7/*Alcaligenes faecalis* S-7 [174]	ACA	Whole cells	7.7	94
Bacillus sphaericus N-7/*Paracoccus denitrificans* [175]	ACA/asp	Whole cells	75.9	92
Corynebacterium sp. C-23/*Paracoccus denitrificans* pFPr-1 [177]	ACA/asp	Co-immobilized cells	24.3	98
Pseudomonas denitrificans [178]	PLA	Whole cells	13.5	90

[a] during batch production
[b] during continuous production
PPA, phenylpyruvic acid; asp, aspartic acid; glu, glutamic acid; PLA, DL-phenyllactic acid; ACA: acetamidocinnamic acid

low ammonia and relatively high isoleucine concentrations resulted in low valine and high phenylalanine production, with no need for an external amino donor [167]. Resting cells of a genetically modified strain of *E. coli* B Z1196 (overexpressing *tyrB*) were capable of producing L-phenylalanine from phenylpyruvate to a final concentration of 28.5 g l^{-1}, with a molar conversion yield of 95% [168, 169]. Extensive screening for phenylpyruvate transforming capacity of microorganisms was performed by Evans et al. [170]. Especially *Pseudomonas* species were found to be superior producers of L-phenylalanine. Highest phenylalanine accumulation and productivity was observed with *Pseudomonas fluorescens* ATCC 11250. AAT activities were highest following growth of the organism in media with L- or D-phenylalanine. The presence of aspartate was essential for efficient conversion; its absence resulted in a 96% reduction of product formation. Highest phenylalanine productivity was observed at 37 °C and strong alkaline pH values (pH 10–12). Immobilization of cells in calcium alginate subsequently allowed continuous production of L-phenylalanine in a packed bed bioreactor, at concentrations greater than 15 g l^{-1} over a period of 60 d. Nakamichi et al. [171] isolated PFP resistant mutants of *Paracoccus denitrificans* which expressed significantly higher levels of aminotransferase activity. High concentrations of L-aspartate and phenylpyruvate were found to inhibit enzyme activity and these substrates therefore were added to the reaction medium at appropriate time intervals. Batch incubations of intact cells resulted in L-phenylalanine production to a level of 122 g l^{-1} with a conversion yield of 92.5% over a period of 72 h. $\varkappa$-Carrageenan immobilized cells of the organism were subsequently used for continuous production of L-phenylalanine at a concentration of 22.3 g l^{-1} and a conversion yield of 90% over a period of 30 d. An alternative approach was chosen by Matsunaga et al. [140], using co-immobilized *Clostridium butyricum*-alanine dehydrogenase-*Micrococcus luteus* as biocatalysts for the production of L-phenylalanine from phenylpyruvate, pyruvate and NH_4Cl under a high hydrogen pressure. In this system pyruvate is transformed into alanine by the alanine dehydrogenase and alanine acts as amino-group donor for phenylalanine synthesis from phenylpyruvate. This resulted in accumulation of 12 g l^{-1} L-phenylalanine and a conversion yield of 80%. Calton et al. [172] observed maximal L-phenylalanine production with *E. coli* B 11303, possessing high L-aspartate: phenylpyruvate aminotransferase activity. Incubation of polyazetidine immobilized cells in a solution of phenylpyruvate plus aspartate resulted in a maximum L-phenylalanine titer of 30 g l^{-1} with a molar yield of 98%. Excellent biocatalyst stability was observed over a period of 100 d. Oxaloacetate produced from aspartate was rapidly decarboxylated to pyruvate under the experimental conditions, thereby driving the reaction to completion [172]. Similar observations were made with an inducible L-aspartate: phenylpyruvate aminotransferase from *P. putida* [163]. Purified enzyme was used to study conditions required for optimal conversion of phenylpyruvate into L-phenylalanine. Decarboxylation of oxaloacetate produced from aspartate turned out to be the limiting step and the conversion rate was dramatically enhanced by including malate dehydrogenase. A comparative study on L-phenylalanine productivity was subsequently reported by Ziehr et al. [173], using whole cells of *P. putida* and purified L-aspartate- phenylpyruvate aminotransferase. Chitosan immobilized whole cells, incubated in a continuously stirred tank reactor, and purified enzyme, incubated behind an ultrafiltration membrane in an enzyme-membrane reactor, produced 6.6 and 9.4 g l^{-1} L-phenylalanine, respectively, from phenylpyruvate and L-aspartate. The productivity of the free enzyme system was 3 times higher, reflecting the higher substrate diffusion rate of this system. The immobilized cells on the other hand displayed better biocatalyst stability. The results suggest that the economy of the process could still be improved by stimulating the decomposition of oxalacetate.

Nakamichi et al. [174] reported the isolation of bacteria with ACA acylase activity. Isolates able to catalyze conversion of ACA via phenylpyruvate into L-phenylalanine were identified as *B. sphaericus* N-7 and *Alcaligenes faecalis* S-7. Conversion of phenylpyruvate into L-phenylalanine was attributed to aminotransferase activity. Further improvement of this whole cell bioconversion process was achieved by combining *B. sphaericus* N-7 (containing highest ACA acylase activity) and *Pa. denitrificans* IFO 12442 (the highest aminotransferase producer) [175]. Among various amino acids tested, maximal conversion of phenylpyruvate was observed with L-aspartic acid as aminodonor, which was attributed to decomposition of the reaction product oxaloacetate. Incubation of both organisms in a 1:1 activity ratio of the two enzymes resulted in production of 75.9 g l^{-1} of L-phenylalanine in 72 h with a conversion yield of 92%. Further work involved the isolation of ACA acylase-hyperproducing, glucose catabolite repression resistant, mutants of *Corynebacterium* sp. S-5 [176]. Continuous production of L-phenylalanine from ACA and L-aspartic acid was subsequently studied with $\varkappa$-carrageenan immobilized cells of mutant C-23 [176], with a 12-fold higher ACA acylase activity, and *Pa. denitrificans* mutant pFPr-1 [171], with increased AAT activity [177]. Co-immobilization of cells was clearly superior to mixtures of immobilized cells, and a 98% molar conversion of ACA was obtained.

Synthesis of L-phenylalanine from D,L-phenyllactate by whole cells of *P. denitrificans* IAM-1426 was reported by Wada [178]. A dehydrogenase was responsible for the conversion of phenyllactate into phenylpyruvate, followed by L-phenylalanine synthesis via aminotransferase activity.

5 Conclusions

The development of microbial and enzymatic processes for the stereospecific synthesis of L-phenylalanine has become the focus of attention in recent years. Also enzymatic processes (e.g., with thermolysin) [101] for the synthesis of α-aspartame have been commercialized. Economic aspects of both microbial and enzymatic processes for L-phenylalanine production have been reviewed by Jones et al. [179].

The application of enzymatic transformations (with whole cells or enzyme preparations) of chemically synthesized achiral precursors is dependent on the cost of precursors and enzymes, the rate of product formation and the overall conversion efficiency of the process. Additional problems are unfavourable reaction equilibria (AAT, PAL), the need for cofactor regeneration (PheDH) and enzyme instability in general. While the search for cheaper feedstocks and more stable enzymes is continuing, excellent progress has been made already in the optimization of conversion yields with the enzymes indicated above.

The development of microbial production processes for L-phenylalanine is equally sensitive to the cost of feedstocks (sugars) and requires extensive manipulation of the production strain to achieve high levels of L-phenylalanine. Traditionally, this involves the isolation of (poly)auxotrophic and amino acid analog resistant strains in order to delete mechanisms involved in controlling L-phenylalanine synthesis via the branched pathway for aromatic amino acids. When following this approach, considerable L-phenylalanine production yields have been achieved especially with *E. coli*, *Corynebacterium* and *Brevibacterium* strains. Although mostly reflected in the patent literature only [180–183], the availability and application of recombinant DNA techniques for strain construction undoubtedly will allow further yield improvement with these organisms. It has to be expected, however, that the intracellular synthesis of precursors for aromatic amino acids ultimately will become ratelimiting for the overall process. This necessitates the identification and, if possible deletion, of alternative processes for PEP (and E4P) utilization (PEP-dependent glucose: PTS transport systems, pyruvate kinase, etc.). Flux control analysis of the pathways involved in L-phenylalanine production appears a valuable approach at this stage [184–186].

Molasses and glucose are most frequently used as microbial growth substrates for L-phenylalanine production. Various alternative feedstocks have been suggested, for instance starch, methanol and *n*-alkanes, which are relatively cheap and abundantly available. Of especial interest appear those feedstocks which as single substrates, or in mixtures with molasses, may serve to increase the carbon flow via biochemical pathways involving precursors for L-phenylalanine synthesis as intermediates. One example is utilization of methanol via the RuMP pathway, involving PEP and E4P [6, 10, 11].

6 Acknowledgements

The authors acknowledge the support by the Foundation for Fundamental Biological Research (BION) which is subsidized by the Netherlands Scientific Organization (NWO). Thanks are due to Prof. W. Harder for valuable discussions.

7 Abbreviations

AAT, aromatic amino acid aminotransferase; ACA, acetamidocinnamic acid; 3-AT, 3-aminotyrosine; DAHP, 3-deoxy-D-arabino-heptulosonate-7-phosphate; E4P, erythrose-4-phosphate; 5-MT, 5-methyl-DL-tryptophan; PAL, L-phenylalanine ammonia-lyase; PEP, phosphoenolpyruvate; PEG, polyethylene glycol; MFP, *m*-fluoro-DL-phenylalanine; OFP, *o*-fluoro-DL-phenylalanine; PFP, *p*-fluoro-DL-phenylalanine; PheDH, L-phenylalanine dehydrogenase; PTS, phosphotransferase system; RuMP, ribulose monophosphate; β-3-TA, β-3-thienyl-DL-alanine.

8 References

1. Crosby GA (1976) CRC Crit. Rev. Food Sci. 7: 297
2. Belitz HD (1986) Aspartam. In: Präve P (ed) Jahrbuch Biotechnologie. Carl Hanser Verlag, München, pp 383–397
3. Klausner A (1985) Bio/Technology 3: 301
4. Atkinson DE (1977) Cellular Energy Metabolism and its Regulation. Academic Press, New York
5. Bloom FR, Kretschmer PJ (1983) Effects of genetic engineering of microorganisms on the future production of amino acids from a variety of carbon sources. In: Wise DL (ed) Organic Chemicals from Biomass. Benjamin/Cummings Publ. Co., London, pp 145—171
6. Dijkhuizen L, Hansen TA, Harder W (1985) Tr. Biotechnol. 3: 262
7. Minoda Y (1986) Raw materials for amino acid fermentation. In: Aida K, Chibata I, Nakayama K, Takinami K, Yamada H (eds) Biotechnology of Amino Acid Production. Kodansha Ltd, Tokyo, Elsevier Science Publ., Amsterdam, pp 51–66 (Progress in Industrial Microbiology, vol 24)
8. Suzuki M, Berglund A, Unden A, Heden C (1977) J. Ferment. Technol. 55: 466
9. Shetty K, Crawford DL, Pometto AL (1986) Appl. Env. Microbiol. 52: 637
10. Morinaga Y, Hirose Y (1984) Production of metabolites by methylotrophs. In: Hou CT (ed) Methylotrophs: Microbiology, Biochemistry, and Genetics. CRC Press, Boca Raton, pp 107—118
11. de Boer L, Vrijbloed JW, Grobben G, Dijkhuizen L (1989) Arch. Microbiol. 151: 319
12. Gibson F, Pittard J (1968) Bacteriol. Rev. 32: 465
13. Umbarger HE (1978) Ann. Rev. Biochem. 47: 533
14. Shiio I (1986) Tryptophan, phenylalanine, and tyrosine. In: Aida K, Chibata I, Nakayama K, Takinami K, Yamada H (eds) Biotechnology of Amino Acid Production. Kodansha Ltd, Tokyo, Elsevier Science Publ., Amsterdam pp 188–206 (Progress in Industrial Microbiology, vol 24)
15. Herrmann KM (1983) The common aromatic biosynthetic pathway. In: Herrmann KM, Somerville RL (eds) Amino Acids: Biosynthesis and Genetic Regulation. Addison-Wesley, London, pp 301–322
16. Garner C, Herrmann KM (1983) Biosynthesis of phenylalanine. In: Herrmann KM, Somerville RL (eds) Amino Acids: Biosynthesis and Genetic Regulation. Addison-Wesley, London, pp 323—338
17. Nyeste L, Pécs M, Sevella B, Holló J (1983) Adv. Biochem. Engin./Biotechnol. 26: 175
18. Enei H, Hirose Y (1985) Phenylalanine. In: Blanch HW, Drew S, Wang DIC (eds) Comprehensive Biotechnology, vol 3. Pergamon Press, Oxford, pp 601–605
19. Nakayama K (1985) Tryptophan. In: Blanch HW, Drew S, Wang DIC (eds) Comprehensive Biotechnology, vol 3. Pergamon Press, Oxford, pp 621–631
20. Haslam E (1974) The Shikimate Pathway. John Wiley, New York
21. Byng S, Kane FF, Jensen RA (1982) CRC Critical Reviews in Microbiology 9: 227
22. Lowe, DA, Westlake DWS (1971) Can. J. Biochem. 49: 448
23. Murphy, MF, Katz E (1980) Can. J. Microbiol. 26: 874
24. Gygax D, Christy M, Ghisalba O, Neusch J (1982) FEMS Microbiol. Lett. 15: 169
25. Tianhui X, Chiao JS (1989) Bioch. Biophys. Act. 991: 1
26. Tianhui X, Chiao JS (1989) Bioch. Biophys. Act. 991: 6

27. Huang L, Montoya AL, Nester EW (1975) J. Biol. Chem. 250: 7675
28. Ely B, Pittard L (1979) J. Bacteriol. 138: 933
29. Weiss U, Edwards JM (1980) The Biosynthesis of Aromatic Compounds. John Wiley, New York
30. Patel N, Pierson DL, Jensen RA (1977) J. Biol. Chem. 252: 5839
31. Hund HK, Keller B, Lingens F (1987) Z. Naturforsch. 42c: 387
32. Keller B, Keller E, Gorisch H, Lingens F (1983) Hoppe-Seyler's Z. Physiol. Chem. 364: 455
33. Ely B, Pittard J (1979) J. Bacteriol. 138: 933
34. McCandliss RJ, Poling MD, Herrmann KM (1978) J. Biol. Chem. 253: 4259
35. Kurahashi O, Noda-Watanabe M, Sato K, Morinaga Y, Enei H (1987) Agric. Biol. Chem. 51: 1785
36. Dopheide TAA, Crewther P, Davidson BE (1972) J. Biol. Chem. 247: 4447
37. Koch GGE, Shaw DC, Gibson F (1971) Biochim. Biophys. Acta 229: 795
38. Nester EW, Jensen RA, Nasser DS (1969) J. Bacteriol. 97: 83
39. Lorence JH, Nester EW (1967) Biochemistry 6: 1541
40. Jensen RA (1969) J. Biol. Chem. 244: 2816
41. Rebello JL, Jensen RA (1970) J. Biol. Chem. 245: 3738
42. Jensen RA, Nester EW (1965) J. Mol. Biol. 12: 468
43. Kane JF, Stenmark SL, Calhoun DH, Jensen RA (1971) J. Biol. Chem. 246: 4308
44. Hagino H, Nakayama K (1974) Agric. Biol. Chem. 38: 2125
45. Hagino H, Nakayama K (1975) Agric. Biol. Chem. 39: 331
46. Hagino H, Nakayama K (1975) Agric. Biol. Chem. 39: 351
47. Hagino H, Nakayama K (1974) Agric. Biol. Chem. 38: 2367
48. Shiio I, Sugimoto S, Miyajima R (1974) J. Biochem. 75: 987
49. Shiio I, Sugimoto S (1981) Agric. Biol. Chem. 45: 2197
50. Sugimoto S, Shiio I (1982) Agric. Biol. Chem. 46: 2711
51. Shiio I, Sugimoto S (1978) J. Biochem. 83: 879
52. Shiio I, Sugimoto S (1976) J. Biochem. 79: 173
53. Sugimoto S, Nakagawa M, Tsuchida T, Shiio I (1973) Agric. Biol. Chem. 37: 2327
54. Shiio I, Ishii K, Yokozeki K (1973) Agric. Biol. Chem. 37: 1991
55. Tsuchida T, Matsui H, Enei H, Yoshinaga F (1974) Japanese Patent Application (Kokai) No. 49-116294
56. Goto E, Ishiwara M, Sakurai S, Enei H, Takinami K (1981) Japanese Patent Application (Kokai) No. 56-64793
57. Shiio I, Sugimoto S, Kawamura K (1988) Agric. Biol. Chem. 52: 2247
58. Hagino H, Nakayama K (1974) Agric. Biol. Chem. 38: 157
59. Ozaki A, Katsumata R, Oka T, Furuya A (1985) Agric. Biol. Chem. 49: 2925
60. Hwang SO, Gil GH, Cho YJ, Kang KR, Lee JH, Bae JC (1985) Appl. Microbiol. Biotechnol. 22: 108
61. Tokoro Y, Oshima K, Okii M, Yamaguchi K, Tanaka K, Kinoshita S, (1970) Agric. Biol. Chem. 34: 1516
62. Tsuchida T, Kubota K, Morinaga Y, Matsui H, Enei H, Yoshinaga F (1987) Agric. Biol. Chem. 51: 2095
63. Choi YJ, Tribe DE (1982) Biotechnol. Lett. 4: 223
64. Akashi K, Shibai H, Hirose Y (1979) J. Ferment. Technol. 57: 321
65. Park SH, Hong KT, You SJ, Lee JH, Bae JC (1984) Korean J. Chem. Eng. 1: 65
66. Gil GH, Kim SR, Bae JC, Lee JH (1985) Enzyme Microb. Technol. 7: 370
67. Förberg C, Häggström L (1987) Appl. Microbiol. Biotechnol. 26: 136
68. Förberg C, Häggström L (1988) J. Biotechnol. 8: 291
69. Sugimoto S, Yabuta M, Kato N, Seki T, Yoshida T, Taguchi H (1987) J. Biotechnol. 5: 237
70. Mori M, Shiio I (1987) Agric. Biol. Chem. 51: 129
71. Ozaki H, Shiio I (1983) Agric. Biol. Chem. 47: 1569
72. Mori M, Shiio I (1984) Agric. Biol. Chem. 48: 1189
73. Mori M, Shiio I (1987) Agric. Biol. Chem. 51: 2671

74. Yokota A, Shiio I (1988) Agric. Biol. Chem. 52: 455
75. Shiio I, Yokota A, Sugimoto S (1987) Agric. Biol. Chem. 51: 2485
76. Sano K, Ito K, Miwa K, Nakamori S (1987) Agric. Biol. Chem. 51: 597
77. Menkel E, Thierbach G, Eggeling L, Sahm H (1989) Appl. Environm. Microbiol. 55: 684
78. Zabriskie DW, Arcuri EJ (1986) Enzyme Microb. Technol. 8: 706
79. Imanaka T (1986) Adv. Biochem. Eng./Biotechnol. 33: 1
80. Schwab H (1988) Adv. Biochem. Eng./Biotechnol. 37: 129
81. Camakaris H, Pittard J (1983) Tyrosine biosynthesis. In: Herrmann KM, Somerville RL (eds) Amino Acids: Biosynthesis and Genetic Regulation. Addison-Wesley, London, pp 339—350
82. Somerville RL (1983) Tryptophan: biosynthesis, regulation, and large-scale production. In: Herrmann KM, Somerville RL (eds) Amino Acids: Biosynthesis and Genetic Regulation. Addison-Wesley, London, pp 351–378
83. Pittard AJ (1987) Biosynthesis of the aromatic amino acids. In: Ingraham JL, Low LB, Magasanik B, Schaechter M, Umbarger HE (eds) *Escherichia coli* and *Salmonella typhimurium*: Cellular and Molecular Biology, vol 2. American Society for Microbiology, Washington, D.C., pp 368–394
84. Hudson GS, Davidson BE (1984) J. Mol. Biol. 180: 1023
85. Ray JM, Yanofsky C, Bauerle R (1988) J. Bacteriol. 170: 5500
86. Beppu T (1986) Application of recombinant DNA technology of breeding of amino-acid-producing strains. In: Aida K, Chibata I, Nakayama K, Takinami K, Yamada H (eds) Biotechnology of Amino Acid Production. Kodansha Ltd, Tokyo, Elsevier Science Publ., Amsterdam, pp 24–35 (Progress in Industrial Microbiology, vol 24)
87. Niederberger P (1989) Amino acid production in microbial eukaryotes and prokaryotes other than coryneforms. In: Baumberg S, Hunter IS, Rhodes PM (eds) Microbial Products: New Approaches. Cambridge University Press, Cambridge, pp 1–24
88. Sano K (1987) Genetic engineering of *Brevibacterium lactofermentum* to breed amino acid producers. In: Neijssel OM, van der Meer RR, Luyben KChAM (eds) Proc. 4th European Congress on Biotechnology. Elsevier Science Publ., Amsterdam, pp 735–747 (Proc. ICEAM Intern. Symp. Amino Acid Fermentations, Demain AL, ed)
89. Katsumata R, Mizukami T, Ozaki A, Kikuchi Y, Kino K, Oka T, Furaya A (1987) Gene cloning in glutamic acid bacteria: the system and its applications. In: Neijssel OM, van der Meer RR, Luyben KChAM (eds) Proc. 4th European Congress on Biotechnology. Elsevier Science Publishers, Amsterdam, pp 767–776 (Proc. ICEAM Intern. Symp. Amino Acid Fermentations, Demain AL, ed)
90. Martın JF, Santamaria R, Sandoval H, Del Real G, Mateos LM, Gil GA, Aguilar A (1987) Bio/Technology 5: 137
91. Martín JF (1989) Molecular genetics of amino acid-producing corynebacteria. In: Baumberg S, Hunter IS, Rhodes PM (eds) Microbial Products: New Approaches. Cambridge University Press, Cambridge, pp 1—59
92. Matsui K, Miwa K, Sano K (1988) Agric. Biol. Chem. 52: 525
93. Folletie MT, Sinskey AJ (1986) J. Bacteriol. 167: 692
94. Matsui K, Miwa K, Sano K (1987) J. Bacteriol. 169: 5330
95. Schmidheini T, Sperisen P, Paravicini G, Hütter R, Braus G (1989) J. Bacteriol. 171: 1245
96. Guroff G, Ito T (1964) J. Biol. Chem. 240: 1175
97. Chandra P, Vining LC (1968) Can. J. Microbiol. 14: 573
98. Friedrich B, Schlegel HG (1972) Arch. Microbiol. 83: 17
99. de Boer L, Harder W, Dijkhuizen L (1988) Arch. Microbiol. 149: 459
100. Yonaha K, Soda K (1986) Adv. Biochem. Eng./Biotechnol. 33: 95
101. Meijer EM, Boesten WHJ, Schoemaker HE, van Balken JAM (1985) Use of biocatalysts in the industrial production of speciality chemicals. In: Tramper J, van der Plas HC, Linko P (eds) Biocatalysts in Organic Syntheses. Elsevier Science Publ., Amsterdam, pp 135–156 (Studies in Organic Chemistry 22)

102. Kamphuis J, Kloosterman M, Schoemaker HE, Boesten WHJ, Meijer EM (1987) Chiral intermediates and applications. In: Neijssel OM, van der Meer RR, Luyben KChAM (eds) Proc. 4th European Congress on Biotechnology. Elsevier Science Publ., Amsterdam, pp 331—348
103. Asai T, Aida K, Oishi K (1959) J. Gen Appl. Microbiol. 5: 150
104. Koukol J, Conn EE (1961) J. Biol. Chem. 236: 2693
105. Young MR, Towers GHN, Neish AC (1966) Can. J. Bot. 44: 341
106. Ogata K, Uchiyama K, Yamada H (1967) Agric. Biol. Chem. 31: 200
107. Ogata K, Uchiyama K, Yamada H (1966) Agric. Biol. Chem. 30: 311
108. Moore K, Subba Rao PV, Towers GHN (1968) Biochem. J. 106: 507
109. Gilbert HJ, Tully M (1982) J. Bacteriol. 150: 498
110. Wick JF, Willis JE (1982) Arch. Biochem. Biophys. 2: 385
111. Sikora LA, Marzluf GA (1982) J. Bacteriol. 150: 1287
112. Onishi N, Yokozeki K, Hirose Y, Kubota K (1987) Agric. Biol. Chem. 51: 291
113. Bezanson GS, Desaty D, Emes AV, Vining LC (1970) Can. J. Microbiol. 16: 147
114. Emes AV, Vining LC (1970) Can. J. Biochem. 48: 613
115. Marconi W, Bartoli F, Gianna R, Morisi F, Spotorno G (1980) Biochimie 62: 575
116. Mushi NY, Kupletskaya MB, Bab'eva IP, Egorov NS (1980) Mikrobiologiya 49: 269
117. Evans CT, Hanna K, Conrad D, Peterson W, Misawa M (1987) Appl. Microbiol. Biotechnol. 25: 406
118. Yamada S, Nabe K, Izuo N, Nakamichi K, Chibata I (1981) Appl. Env. Microbiol. 42: 773
119. Nakamichi K, Nabe K, Yamada S, Chibata I (1983) Eur. J. Appl. Microbiol. Biotechnol. 18: 158
120. Marusich WC, Jensen RA, Zamir LO (1981) J. Bacteriol. 146: 1013
121. Kane JF, Fiske MJ (1985) J. Bacteriol. 161: 963
122. Gilbert HJ, Stephenson JR, Tully M (1983) J. Bacteriol. 153: 1147
123. Gilbert HJ, Clarke IN, Gibson RK, Stephenson JR, Tully M (1985) J. Bacteriol. 161: 314
124. Anson JG, Gilbert HJ, Oram JD, Minton NP (1987) Gene 58: 189
125. Evans CT, Payne C, Conrad D, Hanna K, Misawa M (1987) Can. J. Microbiol. 33: 636
126. Evans CT, Conrad D, Hanna K, Peterson W, Choma C, Misawa M (1987) Appl. Microbiol. Biotechnol. 25: 399
127. Evans CT, Choma C, Peterson W, Misawa M (1987) J. of Ind. Microbiol. 2: 53
128. Evans CT, Hanna K, Payne C, Conrad D, Misawa M (1987) Enzyme Microb. Technol. 9: 417
129. Evans CT, Choma C, Peterson W, Misawa M (1987) Biotechnol. Bioeng. 30: 1067
130. Orndorff SA, Constantino N, Stewart D, Durham DR (1988) Appl. Env. Microbiol. 54: 996
131. Hamilton BK, Hsiao H, Swann WE, Anderson DM, Delente JJ (1985) Tr. Biotechnol. 3: 64
132. Hummel W, Weiss N, Kula MR (1984) Arch. Microbiol. 137: 47
133. Chibata I, Tosa T, Sano R (1965) Appl. Microbiol. 13: 618
134. Hummel W, Schütte H, Schmidt E, Wandrey C, Kula MR (1987) Appl. Microbiol. Biotechnol. 26: 409
135. Asano Y, Nakazawa A (1985) Agric. Biol. Chem. 49: 3631
136. Asano Y, Nakazawa A (1987) Agric. Biol. Chem. 51: 2035
137. Asano Y, Nakazawa A, Endo K (1987) J. Biol. Chem. 262: 10346
138. Asano Y, Nakazawa A, Endo K, Hibino Y, Ohmori M, Numao N, Kondo K (1987) Eur. J. Biochem. 168: 153
139. Evans CT, Bellamy W, Gleeson M, Aoki H, Hanna K, Peterson W, Conrad D, Masawa M (1987) Bio/Technology 5: 818
140. Matsunaga T, Higashijima M, Nakatsugawa H, Nishimura S, Kitamura T, Tsuji M, Kawaguchi T (1987) Appl. Microbiol. Biotechnol. 27: 11
141. Misono H, Yonezawa J, Nagata S, Nagasaki S (1987) J. Bacteriol. 171: 30

142. Ohshima T, Sugimoto H, Soda K (1988) Anal. Lett. 21: 2205
143. de Boer L, van Rijssel M, Euverink GJ, Dijkhuizen L (1989) Arch. Microbiol. 153: 12
144. Hummel W, Schmidt E, Wandrey C, Kula MR (1986) Appl. Microbiol. Biotechnol. 25: 175
145. Kula MR, Hummel W, Schütte H, Leuchtenberger W (1984) Offenlegungsschrift DE 33 07 095 A1, Deutsches patentamt
146. Leuchtenberger W, Kula MR, Hummel W, Schütte H (1986) Offenlegungsschrift DE 3446304 A1, Deutsches Patentamt
147. Hummel W, Schütte H, Kula MR (1988) Anal. Biochem. 10: 397
148. Hummel W, Schütte H, Schmidt E, Kula MR (1987) Appl. Microbiol. Biotechnol. 27: 283
149. Schmidt E, Vasic-Racki D, Wandrey C (1987) Appl. Microbiol. Biotechnol. 26: 42
150. Matsunaga T, Higashijima M, Sulaswatty A, Nishimura S (1988) Biotechnol. Bioeng. 31: 834
151. Ohshima T, Soda K (1989) Tr. Biotechnol. 7: 210
152. Campagna R, Bückmann AF (1987) Appl. Microbiol. Biotechnol. 26: 417
153. Nalbach U, Schiemenz H, Stamm WW, Hummel W, Kula MR (1988) Anal. Chim. Acta 213: 55
154. Okazaki N, Hibino Y, Asano Y, Ohmori M, Numao N, Kondo K (1988) Gene 63: 337
155. Asano Y, Endo K, Nakazawa A, Hibino Y, Okazaki N, Ohmori M, Numao N, Kondo K (1987) Agric. Biol. Chem. 51: 2621
156. Hummel W, Schütte H, Kula MR (1985) Appl. Microbiol. Biotechnol. 21: 7
157. Schütte H, Hummel W, Kula MR (1984) Appl. Microbiol. Biotechnol. 19: 167
158. Jensen RA, Calhoun DH (1981) CRC Crit. Rev. Microbiol. 8: 229
159. Lee CW, Desmazeaud MJ (1985) Arch. Microbiol. 140: 331
160. Lee CW, Desmazeaud MJ (1985) J. Gen. Microbiol. 131: 459
161. Lee CW, Desmazeaud ML (1986) FEMS Microbiol. Lett. 33: 95
162. Kishore G, Sugumaran M, Vaidyanathan CS (1976) J. Bacteriol. 128: 182
163. Ziehr H, Kula MR (1985) J. Biotechnol. 3: 19
164. Lee CW, Lucas S, Desmazeaud MJ (1985) FEMS Microbiol. Lett. 26: 201
165. Schutten I, Harder W, Dijkhuizen L (1987) Appl. Microbiol. Biotechnol. 27: 292
166. Kitai A, Kitamura J, Miyachi M (1962) Hakko To Taisha 5: 61
167. Bulot E, Cooney CL (1985) Biotechnol. Lett. 7: 93
168. Then J, Doherty A, Neatherway H, Marquardt R, Deger HM, Voelskow H, Wöhner G, Präve P (1987) Biotechnol. Lett. 9: 680
169. Robinson M, Marquardt R, Then J, McChesney J, Neatherway H, Wöhner G, Deger HM, Präve P (1987) Biotechnol. Lett. 9: 673
170. Evans CT, Peterson W, Choma C, Misawa M (1987) Appl. Microbiol. Biotechnol. 26: 305
171. Nakamichi K, Nabe K, Nishida Y, Tosa T (1989) Appl. Microbiol. Biotechnol. 30: 243
172. Calton GJ, Wood LL, Updike MH, Lantz L, Hamman JP (1986) Bio/Technology 4: 317
173. Ziehr H, Kula MR, Schmidt E, Wandrey C, Klein J (1987) Biotechnol. Bioeng. 9: 482
174. Nakamichi K, Nabe K, Yamada S, Tosa T, Chibata I (1984) Appl. Microbiol. Biotechnol. 19: 100
175. Nakamichi K, Nishida Y, Nabe K, Tosa T (1985) Appl. Biochem. and Biotechnol. 11: 367
176. Nakamichi K, Nabe K, Tosa T (1986) J. Biotechnol. 4: 293
177. Nishida Y, Nakamichi K, Nabe K, Tosa T (1987) Enzyme Microb. Technol. 9: 479
178. Wada H (1974) J. Agric. Chem. Soc. Jap. 48: 351
179. Jones JL, Fong WS, Hall P, Cometta S (1988) Chemtech (May): 304
180. Lawlis VB, Rastetter WH, Snedecor BR (1985) European Patent Application 84305532.8
181. Backman KC (1985) European Patent Application 84306841.2
182. Edwards MR, Taylor PP, Hunter MG, Fotheringham IG (1987) International Patent Application PCT/US86/01353

Table 1. Survey of important processes for the synthesis of C-5 substituted hydantoins (compiled from [19])

Condensation using	Reactants	
Carboxyl-compounds	α-Amino acids α-Amino acid amides α-Amino nitriles	+ Potassium cyanate (KOCN)
	α-Amino acids α-Hydroxy acids α-Hydroxy nitriles (cyanohydrins)	+ Urea
	α-Amino acid esters α-Amino amides α-Amino nitriles	+ Alkylchloroformates ($ClC(=O)-O-Alk$) or phosgene ($COCl_2$)
	α-Amino nitriles + CO_2	
Carbonyl-compounds (Bucherer-Bergs synthesis)	Aldehyde + KCN + $(NH_4)_2CO_3$ (ketone)	
C-5 substitution with carbonyl-compounds	Aldehyde, ketone or related compounds + hydantoin; reducing agent	

2.2 Methods of Chemical Synthesis of C-5 Substituted Hydantoins

2.2.1 Classification

The main groups of chemical reactions resulting in C-5 substituted hydantoins are listed in Table 1. In general, either the Bucherer-Bergs synthesis from carbonyl compounds (Fig. 3) or the condensation of aldehydes with hydantoin (Fig. 4) are the preferred reactions. The applicability of either method depends on the nature of the C-5 residue desired and on the availability of precursors to allow appropriate and cost-effective introduction of the 5-substituent.

The carbonyl precursors used for the Bucherer-Bergs synthesis must contain an additional methylene-group as compared with those required for the second way (Figs. 3 and 4). Furthermore, synthetic strategies for the construction of the C-5 substituent may predetermine the method of choice. For example, an indolylmethylene side chain may be introduced into the hydantoin ring by substitution of the C-5 position with indole-3-aldehyde, but also by the Bucherer-Bergs synthesis using phenylhydrazine as precursor of the indole ring [15–18].

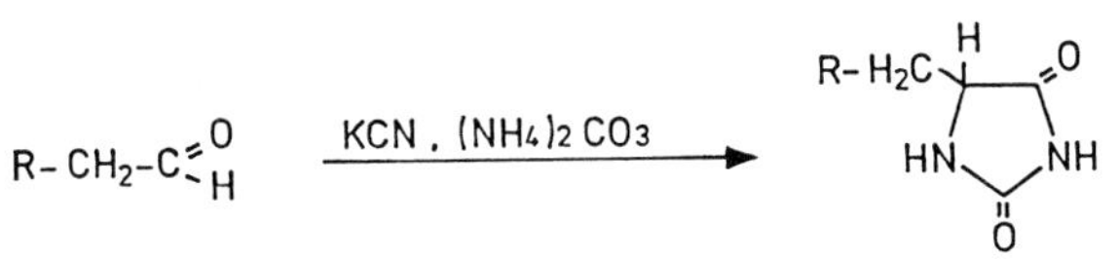

Fig. 3. Bucherer-Bergs synthesis of C-5 substituted hydantoins from carbonyl compounds

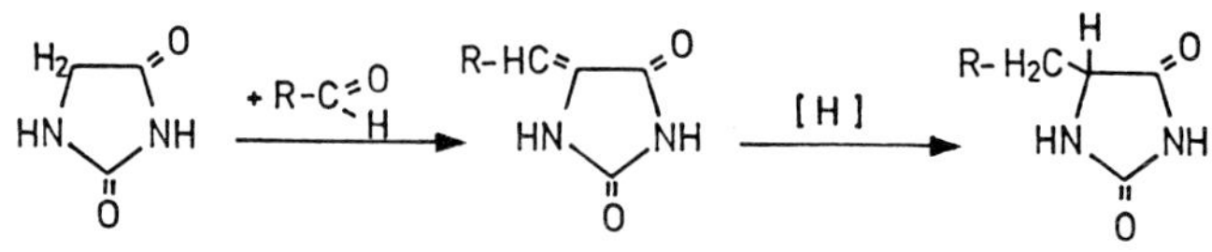

Fig. 4. C-5 substitution of hydantoin with carbonyl compounds

2.2.2 Ring Condensation Starting from Carboxyl- or Related Compounds

A frequently used method for the synthesis of diverse hydantoins starts with the conversion of α-amino acids to the corresponding hydantoic acids (*N*-carbamoyl-amino acids) obtained through the action of potassium cyanate on the precursor, and subsequent condensation of the intermediate with hydrochloric acid under heating. Thus, the α-carbon atom of the amino acids forms the C-5 position of the hydantoin ring. It has occasionally been found advantageous to use instead of α-amino acids their amides or nitriles; especially the latter could efficiently be reacted with carbon dioxide under pressure.

While most of these reactions were carried out under acidic conditions with hydrochloric or glacial acetic acid, it has been shown that hydantoic acids, and consecutively 5-substituted hydantoins, could also be obtained in good yields by boiling the amino acids in barium hydroxide solution with an excess of urea. Furthermore, conversions of α-hydroxy acids or α-hydroxy nitriles (cyanohydrins) with urea were also successfully performed: in this case, ring closure of the resulting *N*-carbamoyl-intermediates were catalyzed by diluted mineral acids. In this way, e.g. 5-phenylhydantoin was prepared from benzaldehyde cyanohydrin.

Starting from α-amino acid amides, both amino groups available can readily be bridged to form the hydantoin ring through the action of carboxyl-chlorides, e.g. alkyl-chloroformates and phosgene, as coupling agents. Thus a series of 5-alkylated and arylated hydantoins were synthesized.

2.2.3 Ring Condensation Starting from Carbonyl- or Related Compounds (Bucherer-Bergs Synthesis)

The "Bucherer Bergs" synthesis was developed in the 1920s and 1930s and since then has found wide-spread application. Aldehydes or ketones are treated with potassium cyanide and ammonium carbonate under mild conditions (Fig. 3). With the exception of formaldehyde, this prodedure was found to be effective for many types of carbonyl compounds, certain unsaturated aldehydes, certain hydroxy- and nitro-aryl aldehydes, and some ketones. Thus, hydantoin itself cannot be obtained by use of formaldehyde, but may be prepared from aminoacetonitrile and ammonium carbonate in aqueous solution at elevated temperature under pressure.

Many processes for the synthesis of C-5 substituted hydantoins are based on this versatile reaction, frequently modified and adapted to the specific problem. Especially in large scale conversions, the low molecular, inorganic components are often applied in their gaseous state as hydrocyanic acid, ammonia, and carbondioxide.

2.2.4 Introduction of C-5 Substituents into the Hydantoin Ring

It has often been found convenient to substitute the C-5 position of the hydantoin ring with aldehydes or related compounds. The condensation reaction with the active 5-methylene group of the hydantoin ring leads to an exocyclic C=C double bond (Fig. 4). It may be carried out in acidic media, but amines like pyridine, piperidine, and diethylamine are also effective as condensing agents.

While a great number of aromatic and heterocyclic aldehydes (among them aldehydes of furan, thiophene, pyrrole, pyridine, quinoline, indole, and imidazole) are reactive, aryl-substituted aliphatic aldehydes, e.g. phenylacetaldehyde and *p*-methoxyphenylacetaldehyde, will not condense with hydantoin. The condensation of some aliphatic and also α,β-unsaturated aldehydes like cinnamaldehyde was, however, reported.

The C-5 unsaturated hydantoin derivatives may be reduced by any of a variety of common reducing agents. However, cleavage of the hydantoin ring or of certain substituents as well as their reduction may accompany this reaction.

2.3 Properties of Hydantoins

Keto-enol tautomerism is a typical feature of the hydantoin structure. Under neutral conditions, the keto-form is dominant; in alkaline solution, enolization between the 4- and 5-positions can occur as has been concluded from the fact that optically pure hydantoins readily racemize, even with both nitrogen atoms substituted. This phenomenon is of practical relevance for the complete conversion of racemic hydantoin derivatives to chiral α-amino acids. It has been suggested, though, that racemization is accelerated by appropriate microbial enzyme systems.

The rate of spontaneous racemization is very much influenced by the electronic properties of the C-5 substituent. Substituents exerting an electronegative inductive effect will stabilize the enolate structure because electron density at C-5 is lowered thus favouring the release of the proton at C-5 (comp. Fig. 2). Therefore hydantoins carrying e.g. a carboxyl function on an alkyl side chain like 5-(2′-carboxyethylene)hydantoins will readily racemize often within minutes. On the other hand it may take hours to racemize merely alkylated hydantoins.

The same applies for 5-aryl-alkylated hydantoins, e.g. 5-benzyl- or 5-indolylmethylenehydantoin. In contrast direct arylation at C-5, e.g. in 5-phenylhydantoin again favours spontaneous racemization, since the C-4 enolate structure forms a conjugated mesomeric system. This facilitates electron dislocation over the 5-aryl-substituent and the hydantoin ring.

Hydantoins are fairly acidic (comparable with phenol), resulting from the N—H group in position 3; acidity is lost when N-3 is substituted. All hydantoins are more or less unstable in the presence of alkali. Substitution in 1- and 5-position increases, but in N-3 position decreases stability. Unsaturation at the C-5 atom has been stated to further stabilize the hydantoin ring towards alkaline hydrolysis. On the other hand, substituents on both nitrogen atoms and the C-5 atom appear to endow remarkable stability towards acid hydrolysis.

3 Distribution of Hydantoin Hydrolyzing Activity in Nature

3.1 Screening Methods

Many microorganisms are able to grow on D,L-5-monosubstituted hydantoins as sole C- and/or N-source in a mineral salt medium. Several methods are proposed for the isolation of microorganisms which are able to hydrolyze these substrates either to *N*-carbamoyl-D- or -L-amino acids or to the corresponding D- or L-amino acids.

Starting from stock cultures the microorganisms are cultivated in the specific media. They can be tested for hydantoin hydrolyzing activity as growing cells on agar plates using specially designed overlay assays [20, 21], as growing cells in liquid culture, as resting cells in a simple buffer solution (e.g. [8, 10, 22–27]) or as crude enzyme extracts [28] with the specific hydantoin as substrate. In order to enrich hydantoin hydrolyzing microorganisms from soil or water samples mineral salt media, which contain the requested hydantoin as sole C- and/or N-source, are inoculated with samples from the respective environment [23, 26, 27, 29, 30, 38]. The test tubes or shake-flasks are then incubated aerobically until growth of microorganisms is recognized. After repeated transfers, the cells can be isolated on agar plates [23, 25, 29] and tested for purity. For selecting microorganisms with high *N*-carbamoyl-amino acid amidohydrolase activity, the respective *N*-carbamoyl-amino acids should be used as C- and/or N-source [31].

3.2 Distribution in Nature

In 1932, Sobotka [32] reported on the enzymatic cleavage of the levo-rotatory form of 5-phenyl-5-ethylhydantoin by growing cells of an *Aspergillus* sp. In 1946, Bernheim and Bernheim [33] detected the ability of liver suspensions of omnivorous animals to rapidly hydrolyze hydantoin, while extracts of other organs or those of herbivora were not active. In 1947, they were able to detect this activity in various plant seeds [34]. Investigation on the microbial cleavage of various D,L-5-monosubstituted hydantoins showed that the ability to hydrolyze these substrates was wide spread in nature, and that D- as well as L-selective enzymes were existing. These so-called hydantoinases were inducible in all cases.

3.2.1 D-Specific Hydantoin Cleavage

Using D,L-5-(2′-methylthioethylene)hydantoin as screening substrate Yamada et al. [8] found hydantoinase activity in all bacteria, actinomycetes, yeasts, and moulds tested. *Pseudomonas striata* was the best strain, unsubstituted hydantoin and 5-methylhydantoin were the best inducers of enzyme formation. Wanru [22] had similar results, using 5-methylhydantoin as screening substrate. He found that especially strains of the genus *Pseudomonas* possessed good D-hydantoinase activities.

While many microorganisms and extracts of mammalian and plant cells [35] catalyze the formation of *N*-carbamoyl-D-amino acids from the corresponding D,L-5-monosubstituted hydantoins, there are only few examples for microorganisms able to form the free D-amino acids. Yokozeki et al. [23] observed that of more than 430 bacteria, actinomycetes, and yeasts tested, only two *Achromobacter* sp. were able to form D-*p*-hydroxyphenylglycine from the corresponding hydantoin. In a second screening of about 2,000 strains isolated from soil with D,L-5-*p*-hydroxyphenylhydantoin as screening substrate, only two microorganisms were able to produce D-*p*-hydroxyphenylglycine. Möller et al. [27] demonstrated that only twelve of 82 strains tested could form free D-serine from D,L-hydroxymethylenehydantoin, when they had grown with this substrate as sole C-source.

3.2.2 L-Selective Hydantoin Cleavage

A screening for microorganisms producing L-glutamic acid from D,L-5-(2′-carboxyethylene)hydantoin showed that many bacteria were able to perform this L-specific reaction, whereby *Bacillus brevis* ATCC 8185 had the highest activity [36]. Yamada et al. [49] were able to isolate a *Bacillus coagulans* able to form L-methionine from D,L-5-(2′-methylthioethylene)hydantoin. Recently Klages et al. [37] isolated a *Nocardia* sp. DSM 3306 able to form L-valine from D,L-5-isopropylenehydantoin as well as Yamashiro et al. [38] a *Bacillus brevis* AJ-12299 able to form L-leucine from D,L-5-isobutylenehydantoin by using the latter as screening substrate. Of about 2,300 strains tested, only one produced L-valine (see Sect. 5.1.3) [38]. It would be of interest whether L-hydantoinase and L-*N*-carbamoyl-amino acid amidohydrolase activities of these microorganisms are identical to those of *Bacillus brevis* ATCC 8185 described by Tsugawa et al. [36].

In contrast to aliphatic amino acids, formation of L-tryptophan from D,L-5-indolylmethylenehydantoin and L-phenylalanine from D,L-5-benzylhydantoin was only observed with bacteria of the genera *Flavobacterium* [25, 26, 42], *Arthrobacter* [21, 30, 39], and *Pseudomonas* [78].

3.3 Biological Function

Hydantoinases belong to the EC 3.5.2 group [43]. Of this EC group, which summarizes enzymes able to hydrolyze cyclic amide bonds, two enzymes hydrolyze naturally occurring hydantoin derivatives: carboxymethylenehydantoinase (EC

Table 2. Hydantoinases and cyclic amidases

Recommended name	Other names	Systematic name	EC-number	Literature
Barbiturase		Barbiturate amidohydrolase	3.5.2.1	[60]
Dihydropyrimidinase	D-Hydantoinase	5,6-Dihydropyrimidine amidohydrolase	3.5.2.2	[8, 11, 22, 27, 28, 33, 34, 38, 55–59]
Dihydro-orotase	Carbamoylaspartic acid dehydrase	L-5,6-Dihydro-orotate amidohydrolase	3.5.2.3	[44, 47, 48]
Carboxymethylhydantoinase		L-5-Carboxymethylhydantoin amidohydrolase	3.5.2.4	[44–46]
Allantoinase		Allantoin amidohydrolase	3.5.2.5	[50–54, 66, 79]
Penicillinase	β-Lactamase, Cephalosporinase	Penicillin amido-β-lactamhydrolase	3.5.2.6	[43]
Imidazolonepropionase		4-Imidazolone-5-propionate amidohydrolase	3.5.2.7	[43]
5-Oxoprolinase	Pyroglutamase	5-Oxo-L-proline amidohydrolase	3.5.2.9	[43]
Creatininase		Creatinine amidohydrolase	3.5.2.10	[43]
		L-5-Carboxyethylhydantoin amidohydrolase		[29, 36, 44]
Indolylmethylhydantoinase		5-Indolylmethylhydantoin amidohydrolase		[26, 30, 40, 41, 63]
1-Methylhydantoinase		1-Methylhydantoin amidohydrolase		[13, 61, 62]

3.5.2.4) and allantoinase (EC 3.5.2.5). All other enzymes have natural substrates containing no hydantoin ring (e.g. 5,6-dihydrouracil is the natural substrate of the so-called "D-hydantoinase", L-5,6-dihydro-orotate the natural substrate of the dihydro-orotase [see below and Fig. 5]). Table 2 gives a survey on the EC 3.5.2 group as well as it specifies other hydantoinases not listed in this group until now.

3.3.1 Carboxymethylenehydantoinase (EC 3.5.2.4)

This enzyme is able to hydrolyze the hydantoin corresponding to L-aspartic acid [44]. It is related to ureidosuccinase (EC 3.5.1.7) [46]. The natural function of these enzymes was postulated to be a side way in the biosynthesis of pyrimidine nucleotides [45]. Carboxymethylenehydantoinase is closely related, but not identical to another cyclic amidase, dihydro-orotase (EC 3.5.2.3) [44]. The latter enzyme was isolated from *Clostridium oroticum* and catalyzes the reversible cleavage of the six-membered ring of dihydro-orotate to ureidosuccinic acid (=*N*-carbamoylaspartic acid) [47, 48].

3.3.2 Allantoinase (EC 3.5.2.5)

Allantoinase (EC 3.5.2.5) is correctly a 5-ureidohydantoinase. It is widely distributed in nature and plays an important role in the degradation of purine nucleotides either in combination with an allantoicase (EC 3.5.3.4) or an allantoate amidohydrolase (EC 3.5.3.9) which hydrolyzes allantoin to urea and glyoxylic acid [50]. The allantoinase is described as inducible and (+)-specific by Lee and Roush [51], Okumura et al. [52] and Trijbels and Vogels [53]. Besides allantoin, some other compounds possessing a free ureido group like *N*-carbamoyl-L-asparagine, *N*-carbamoyl-L-aspartate (the corresponding D-compounds were ineffective), hydantoate (=*N*-carbamoylglycinate) and diureidomethane were found to enhance the synthesis of this enzyme [79]. Investigation on substrate specificity of allantoinases is limited, but e.g. D,L-5-aminohydantoin was shown to be accepted as substrate, although poorly [50]. It would be of interest to test this enzyme concerning the hydrolysis of other D,L-5-monosubstituted hydantoins.

Non-stereospecific allantoin hydrolysis and association of the allantoinase with a cofactor-independent allantoin racemase (EC 5.1.99.3) has been reported [50, 54], enabling some microorganisms to use (—)-allantoin as substrate, too.

3.3.3 Dihydropyrimidinase, 5,6-Dihydropyrimidine Amidohydrolase or D-Hydantoinase (EC 3.5.2.2)

5,6-Dihydropyrimidine is a six-membered ring comparable with 5,6-dihydro-orotate (see Fig. 5). This compound as well as 5,6-dihydrouracil and 5,6-dihydrothymine can be hydrolyzed by the enzyme dihydropyrimidinase (EC 3.5.2.2), which is involved in the degradation of pyrimidine nucleotides. This widely distributed, inducible catabolic enzyme is strictly D-specific in contrast to the L-specific dihydroorotase (EC 3.5.2.3) which is involved in the opposite anabolic pathway (see 3.3.1). The dihydropyrimidinase is identical to the so-called D-hydantoinase [8, 11, 55, 56, 69]. Natural cyclic amides like 5,6-dihydrouracil, uracil and 5,6-dihydrothymine are effective inducers for enzyme formation [8, 22, 24, 27, 28]; so are unsubstituted hydantoin or D,L-5-monosubstituted hydantoins containing a small uncharged 5-substituent (e.g. [8, 22, 23, 27]). The enzyme is able to hydrolyze 5-monosubstituted hydantoins with good activity except those containing charged groups in the side chain (see Table 5) [10, 27, 57, 59, 68]. In some cases, the dihydropyrimidinase is associated with a D-specific D-*N*-carbamoyl-amino acid amidohydrolase, called β-ureidopropionase (EC 3.5.1.6) after its natural substrate *N*-carbamoyl-β-alanine [27, 31, 57, 58], or a hydantoin racemase [39, 80].

The dihydropyrimidinase is closely related with, but not identical to the barbiturase (EC 3.5.2.1) able to hydrolyze barbituric acid and methylbarbituric acid [69]. The only difference of these substrates from uracil and thymine is the presence of a keto-group instead of a methyl- or a hydrogen-group in the 6-position of the ring (see Fig. 5). Barbiturase was detected by Hayaishi and Kornberg [60] in bacteria of the genera *Mycobacterium* and *Corynebacterium* and postulated to catalyze a side way in the degradation of pyrimidines.

Unfortunately there are no reports on the cleavage of D,L-5-monosubstituted hydantoins by this enzyme and on its purification.

3.3.4 Other Hydantoinases

Three other hydantoinases are described in literature, which have not yet been enclosed in the Enzyme Nomenclature [43].

Akamatsu [44] proved by induction experiments that a L-carboxyethylene-hydantoinase described also by Tsugawa et al. [36] and Hassall and Greenberg [29] for the production of L-glutamic acid from the corresponding hydantoin is not identical to the L-carboxymethylenehydantoinase (see 3.3.1). Hassall and Greenberg [29] proposed carboxyethylenehydantoin to be a product of the microbial degradation of L-histidine. The possible identity of this enzyme with the ATP-dependent hydantoinases described by Yamashiro et al. [38, 71] and Klages et al. [37] (see Sect. 5.2) has to be investigated.

Siedel et al. [13] and also Yamada et al. [61, 62] found a new ATP-dependent-1-methylhydantoinase in different bacteria. This inducible enzyme, which also acts on unsubstituted hydantoin and 5-methylhydantoin [13], is involved in the degradation of the cyclic amide creatinine after its deimination in 2-position to 1-methylhydantoin to result in sarcosine (=*N*-methylglycine) via *N*-carbamoyl-sarcosine (=*N*-carbamoyl-*N*-methylglycine) [13, 61]. It is associated with a so-called D-*N*-carbamoylsarcosine hydrolase [62] (see also Sect. 6.4).

Nishida et al. [26], Syldatk et al. [30], Yamashiro et al. [38], and Yokozeki et al. [63] found a new hydantoinase involved in the L-selective cleavage of aryl-alkylenehydantoins, which is inducible only by D,L-5-indolylmethylenehydantoin out of a variety of 5-monosubstituted hydantoins and natural cyclic amides tested [26, 40, 41]. The natural function of this enzyme is not yet known, while the associated *N*-carbamoyl-L-amino acid amidohydrolase was shown by Syldatk et al. [64] to be reactive to *N*-formyl-L-amino acids, too, (see Table 5).

Figure 5 gives an overview on the natural substrates of the enzymes of the EC 3.5.2 group in comparison with the hydantoin ring.

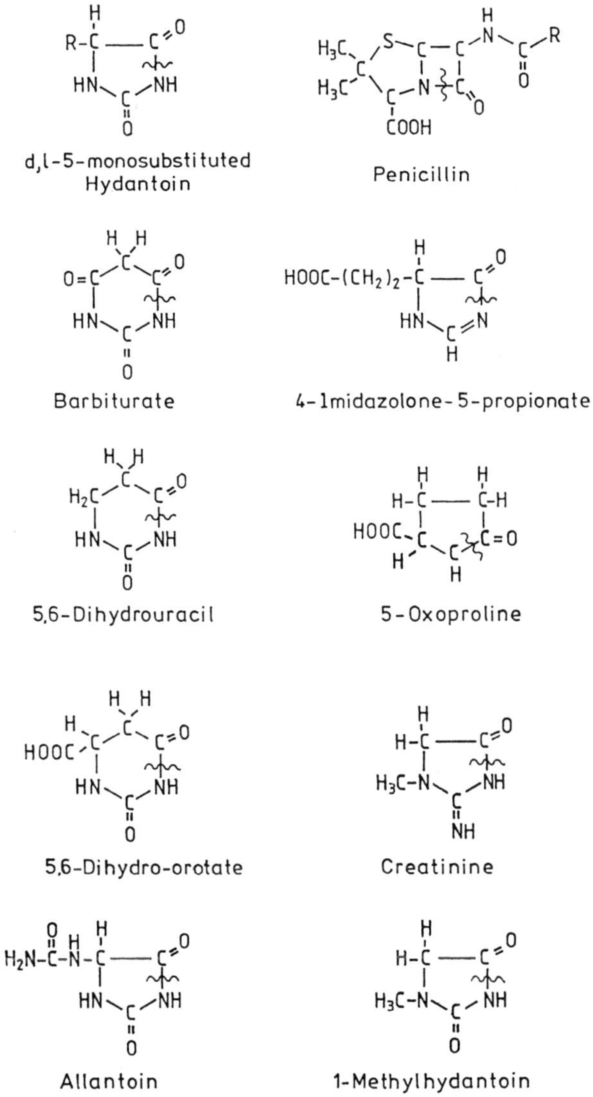

Fig. 5. Structure of D,L-5-monosubstituted hydantoins in comparison with the natural substrates of the EC 3.5.2. group cyclic amidases and 1-methylhydantoinase

3.4 Analytical Methods

3.4.1 Qualitative Detection Methods

After incubating D,L-5-monosubstituted hydantoins in the presence of microbial cells or crude enzyme extracts, the biocatalysts are separated from the reaction solution by centrifugation, filtration, or precipitation. Reaction products in the supernatants can then be separated by paper or thin layer chromatography on silica gel plates (e.g. [8, 10, 23, 25, 27, 31, 41, 58, 61]). *N*-carbamoyl-amino acids and also hydantoins of aromatic amino acids are detected as yellow or red spots by spraying the plates with a solution of *p*-dimethylamino-benzaldehyde and heating subsequently (e.g. [8]). Amino acids are visualized as red or violet spots by spraying the plates or the paper strips with ninhydrin and heating subsequently (e.g. [23, 27, 31]). A frequently used solvent for the separation of hydantoins, *N*-carbamoylamino acids, and amino acids is 1-butanol/acetic acid/water in the ratio (V/V/V) of about 4:1:1 (e.g. [8, 10, 23, 27, 31]). The exact ratio is adjusted to the properties of the compounds to be separated. Recently, Morin et al. [20] offered an elegant variation of this qualitative method for the rapid detection of microorganisms on solid complex media able to hydrolyze dihydrouracil. Positive strains developed a yellow colour around the colonies within 5 to 10 seconds after challenge with 10 μl drops of an acid solution of *p*-dimethylamino-benzaldehyde.

Groß et al. [21] developed an overlay assay for selecting microorganisms able to form L-tryptophan from crystalline D,L-5-indolylmethylenehydantoin on solid complex media. Hydantoinase producing strains formed a clear area around their colonies. A second overlayer with a tryptophan-auxotroph yeast allowed to detect L-tryptophan producing microorganisms.

3.4.2 Quantitative Methods

The *p*-dimethylamino-benzaldehyde as well as the ninhydrin method (see 3.4.1) can be used as photometric methods for the quantitative determination of *N*-carbamoyl-amino acids and amino acids in the reaction supernatants, the latter in combination with an automatic amino acid analyzer [27, 56]. The *p*-dimethyl-amino-benzaldehyde method cannot be applied to aromatic *N*-carbamoyl-amino acids, because the corresponding hydantoins also react. The activity of enriched or purified *N*-carbamoyl-amino acid amidohydrolases can be studied by manometrical methods based on the CO_2-production [35] or by measuring the NH_4^+ released during the reaction [44, 60, 62].

Recent studies apply HPLC-methods on reversed phase columns to the parallel analysis of hydantoins, *N*-carbamoyl-amino acids and amino acids. Aromatic compounds can easily be quantified photometrically at 220 or 280 nm [21, 26, 30, 40, 41], whereas for the detection of 5-alkylhydantoins and their degradation products pre- or post-column derivatization methods have to be applied.

3.4.3 Determination of Enantiomeric Purities

A classical method for quantification of enantiomeric purities is to isolate the products and to measure their optical rotation in comparison with the optically pure D- and L-compounds. In the case of hydantoin hydrolysis this would imply the separation and purification of the optically active hydantoin, *N*-carbamoyl-amino acid and amino acid.

Recently, with the development of chiral columns, HPLC has become an important tool for the detection of the enantiomeric purities. In most cases a foregoing separation of substrate and reaction products is not necessary. By using chiral supports (e.g. copper-D-amino acid complexes), it is possible to determine not only the enantiomeric purity of *N*-carbamoyl-amino acids and amino acids, but also that of 5-monosubstituted hydantoins [57, 63, 65]. Also chiral thin layer chromatography with comparable supports may be used as screening method [28]. Rapid screening can also be carried out by specific bioassays. Sano et al. [25] describe the detection of L-tryptophan by a *Leuconostoc mesenteroides*. Groß et al. [21] used a L-tryptophan-auxotrophic mutant of *Saccharomyces cerevisae*. While these two methods as well as the enzymatic assay for L-glutamic acid described by Tsugawa et al. [36] are specific only for one amino acid, the commercially available enzymes D- and L-amino acid oxidase can be applied to a wide range of amino acids in combination with the photometric ninhydrin method (see above) [27, 28]. To detect the configuration of the *N*-carbamoyl-amino acids produced decarbamoylation with nitrite is applied under mild reaction conditions without a change of their optical rotation [10, 27].

4 D-Selective Hydrolysis of Hydantoin Derivatives

4.1 Microbial Conversion

4.1.1 History

In 1926, Gaebler and Keltch showed that some hydantoin was excreted as hydantoic acid (=*N*-carbamoyl glycine) after injection into dogs [72]. In 1934, Wada measured the production of urea when hydantoin was incubated with milk or other tissue suspensions [73]. Sobotka was the first to report on the microbial cleavage of 5-phenyl-5-ethylhydantoin by an *Aspergillus* sp. [32] (see Sect. 3.2). About 15 years later, Bernheim et al. described the enzymatic hydrolysis of hydantoins with mammalian and plant enzymes [33–35]. Grisolia et al. studied the role of 5,6-dihydropyrimidinase and a so-called *N*-carbamoyl-β-alanine decarbamoylase in the metabolism of pyrimidines [69, 70].

A biotechnological application of this reaction became interesting not before the 1970s together with the growing interest of industry in producing D-amino acids as side chains for semisynthetic penicillins and cephalosporines. At first Cecere et al. [67, 68] successfully tried to use the immobilized calf liver dihydropyrimidinase for the production of *N*-carbamoyl-D-phenylglycine and its *p*-chloro-

and *p*-hydroxy-derivatives from the corresponding chemically synthesized racemic hydantoins (see Sect. 6.2). Various groups in Italy [31, 56, 80], Japan [8–12, 24, 49], France [39], Korea [22, 59] and Germany [6, 37] investigated the D-specific hydantoin cleavage using microorganisms or microbial enzymes.

4.1.2 Microbial Production of *N*-Carbamoyl-D-Amino Acids

Yamada et al. [8] investigated the distribution of hydantoin hydrolyzing activity in microorganisms and found high activity for forming *N*-carbamoyl-D-methionine from D,L-5-(2′-methylthioethylene)hydantoin in various bacteria. Unsubstituted hydantoin and 5-methylhydantoin were found to be the best inducers of hydantoinase activity in *Pseudomonas striata* [8]. Meat extract in combination with glycerol was shown to be advantageous for enzyme formation, which was maximal at the beginning of the stationary growth phase [8]. Investigations on the purified enzyme proved that the D-hydantoinase of *Pseudomonas striata* had a wide substrate specificity even for producing *N*-carbamoyl-derivatives of non-proteinogenic amino acids (see Table 5) [9].

Takahashi et al. had comparable results, using intact cells of *Aerobacter cloacae*, *Corynebacterium sepedonicum* and *Streptomyces griseus* [10]. The results described above led to the development of a process for the production of D-amino acids using resting cells of *Pseudomonas striata* or *Bacillus* sp. immobilized in polyacrylamide and chemical decarbamoylation by treating the *N*-carbamoyl-D-amino acids with equimolar nitrite under acidic conditions [11].

The operational stability of this process was successfully investigated for the cleavage of phenyl-, thienyl-, and *p*-hydroxyphenylhydantoin [11, 12]. Fig. 7 shows the reaction scheme of this process (see 6.2).

Guivarch et al. [39] used a *Pseudomonas* sp. for the production of the *N*-carbamoyl-derivatives of D-methionine, D-phenylalanine, D-cyanopropyleneglycine, D-leucine, and D-valine. In contrast to other authors, they were able to prove enzymatic racemization of L-5-(methylthioethylene)hydantoin.

Wanru [22, 59] proved *Pseudomonas putida* to have the highest activity of dihydropyrimidinase among 275 bacterial strains belonging to 19 genera. Unsubstituted hydantoin, 5-methylhydantoin, 5-phenylhydantoin, uracil, adenine and orotic acid were effective inducers for enzyme formation. Lactic acid in combination with yeast extract gave the best results in the synthesis of dihydropyrimidinase [22]. The substrate specificity of this enzyme [59] was comparable to those described above and is shown in Table 5.

All microorganisms described in this section are only able to form *N*-carbamoyl-D-amino acids from the corresponding hydantoins. To obtain the free D-amino acids the isolation of the intermediary products as well as a second chemical reaction step are necessary [11, 39, 64].

Other publications deal with a D-specific microbial hydantoin cleavage directly leading to the free, optically pure D-amino acids in two enzymatic reactions.

4.1.3 Microbial Production of D-Amino Acids

Olivieri et al. used *Agrobacterium tumefaciens* NRRL B 11291 for the complete conversion of racemic hydantoins to D-amino acids [31, 56]. This microorganism catalyzes the two-step reaction after cultivation in a medium containing 0.2% uracil as sole *N*-cource [31]. In comparison with crude enzyme preparations, intact cells led to a higher ratio of produced D-amino acid versus *N*-carbamoyl-D-amino acid [56]. Permeabilization with toluene was necessary to convert the *N*-carbamoyl-derivatives of the D-amino acids because of the low permeability of the cell membrane to the *N*-carbamoyl-substrates. Compromises had to be found between the optimal reaction conditions for the hydantoinase- and the *N*-carbamoylase-reaction: While intact cells of *Agrobacterium tumefacies* NRRL B 11291 showed a maximal conversion of the hydantoin at 60 °C, maximal D-amino acid formation was obtained at 50 °C.

The repetitive use of *Agrobacterium tumefaciens* cells for the formation of D-*p*-hydroxyphenylglycine was successful [56]. The substrate specificity of the D-hydantoinase of this bacterium was comparable with that of other microbial D-hydantoinases described above (see Table 5). The substrate specificity of the *N*-carbamoyl-D-amino acid amidohydrolase can be seen in Table 6. A use of only the latter enzymatic reaction for the production of D-amino acids was not practical due to the strong inhibition of this enzyme by *N*-carbamoyl-L-amino acids [31]. Results similar to those obtained with *Agrobacterium tumefaciens* (see above) were obtained by using *Arthrobacter crystallopoietes* AM2 [6, 27]. Optimal conditions for resting cells were found to be 55 °C and pH 9.2 [27].

Unsubstituted hydantoin and 5-methylhydantoin were better inducers than dihydrouracil. While cells from the middle of the exponential growth phase only poorly formed free D-amino acids, maximal conversion was obtained with cells harvested at the end of the exponential growth phase. Highest activities were found with unsubstituted hydantoin, 5-methylhydantoin, and 5-hydroxymethylene-hydantoin. Tables 5 and 6 give a survey on the substrate specificities of both enzymes [27].

Yokozeki et al. [23, 24, 57] described the optimal reaction conditions for the production of D-*p*-hydroxyphenylglycine from the corresponding racemic hydantoin by resting cells of *Pseudomonas* sp. AJ-11220 to be pH 8.0 and 43 °C [24]. In contrast to Olivieri et al. [31, 56] and Möller et al. [6, 27], no *N*-carbamoylamino acids were observed as intermediary products [23, 24]. For enzyme formation, a mineral salt medium containing 0.1% yeast extract and 0.05% 5-cyanoethylene-hydantoin as inducer was suitable. Other suitable inducers were 5-methylthio-ethylenehydantoin and 5-(*p*-hydroxyphenyl)hydantoin, while hydantoin, 5-methylhydantoin, uracil, and dihydrouracil caused only slight induction. In contrast to other publications [6, 10, 27, 67, 68], *Pseudomonas* sp. AJ-11220 was able to form D-glutamic acid from the corresponding hydantoin [24]. Purification of the enzymes of *Pseudomonas* sp. AJ-11220 resulted in a D-hydantoinase, a *N*-carbamoyl-L- and a *N*-carbamoyl-D-amino acid amidohydrolase [57]. All three enzymes were strictly D- and L-specific, respectively, and had wide substrate specificities, which can be seen in Tables 5, 8, and 14.

Table 3. Purification of bacterial D-hydantoinases

Source	*Pseudomonas striata*	*Pseudomonas fluorescens*	*Pseudomonas* sp. AJ-11220	*Arthrobacter crystallopoietes*
Reference	[9]	[77]	[57]	[27]
Purification steps:				
1st step	Protamine sulfate	Phenyl-sepharose CL4B	DEAE-Toyopearl	Protamine sulfate
2nd step	Ammonium sulfate	Sephacryl S-400		Ammonium sulfate
3rd step	DEAE-cellulose	Sephacryl S-400		DEAE-cellulose
4th step	Ammonium sulfate	Sephacryl S-400		
5th step	Hydroxylapatite	Preparative gel electrophoresis		
6th step	Sephadex G-200			
7th step	Crystalls			
Yield	3%	1% (after gelfiltration)	63%	
Purification factor	300	0.53 (after gelfiltration)	27	
Remarks	The homogeneous enzyme could be crystallized		Partially purified	Partially purified

4.2 Isolation and Characterization of D-Hydantoinases

4.2.1 Isolation

The enzyme dihydropyrimidine amidohydrolase (EC 3.5.2.2) was purified from animal tissues and bacteria (see Sect. 3.3.3).

Wallach and Grisolia [69] described a procedure of acid and heat treatments followed by several ammonium sulfate and acetone fractionations using water extracts of beef liver acetone powders as enzyme source. The D-hydantoinase could be enriched 200-fold with a yield of 25% and a purity of 80%.

Brooks et al. [75] prepared the D-hydantoinase to homogeneity (13% yield, 24.4-fold purification) from a bovine liver catalase fraction using either an electrophoresis or hydrophobic chromatography.

Bacterial D-hydantoinases were isolated from *Pseudomonas striata* [9], *Pseudomonas fluorescens* DSM 84 [77], *Pseudomonas* sp. AJ-11220 [57], and *Arthrobacter crystallopoietes* AM2 [27]. A survey on the purification procedures is given in Table 3.

A different 1-methylhydantoinase was purified from various microorganisms (*Arthrobacter* sp. DSM 2563, DSM 2564, *Moraxella* sp. DSM 2562, *Micrococcus* sp. DSM 2565, and *Brevibacterium* sp. DSM 2843) using ammonium sulfate and polyethyleneimine precipitations followed by chromatographical steps not further specified [13].

4.2.2 Characterization

Table 4 shows some characteristic properties of the purified D-hydantoinases concerning the ring cleavage (3.3.3). It is remarkable that there is a high degree of conformity of the bacterial enzymes with those isolated from animal tissues.

The dihydropyrimidine amidohydrolase reaction is reversible: Ring cleavage occurs at pH-values around 8.5, while the optimal pH for ring closure is on the neutral or weakly acidic side. But there is no enzymatic conversion of *N*-carbamoyl glycine to hydantoin [69]. Inhibitor studies showed that the enzyme carries an SH-function at the active site. Brooks et al. [75] identified the beef liver D-hydantoinase as a zinc enzyme which contains 4 mol Zn^{2+} per mol active enzyme.

Table 5 gives a survey on the substrate specificities of this group of enzymes. With the exception of the ATP-dependent 1-methylhydantoinase (see 3.3.4) which is not yet totally characterized, the D-hydantoinases can hydrolyze a remarkably broad substrate spectrum. The stereospecificity of the 1-methylhydantoinase has not yet been investigated [13].

4.3 Isolation and Characterization of *N*-Carbamoyl-D-Amino Acid Amidohydrolase

4.3.1 Isolation

Caravaca et al. [70] prepared a 20.4-fold enriched fraction of a D-specific *N*-carbamoyl-amino acid amidohydrolase from rat liver with a yield of 12% by three

Table 4. Properties of D-hydantoinases

Source	Beef liver	Beef liver	*Pseudomonas striata*	*Pseudomonas fluorescens*	*Pseudomonas* sp. AJ. 11220	*Arthrobacter crystallopoietes*
Reference	[69]	[75]	[9]	[77]	[57]	[27]
Molecular weight		226,000 Da	190,000 Da	230,000 Da		
Subunits:						
Number		4		4		
Molecular weight		56,500 Da		60,000 Da		
Optimal temperature			45–55 °C	55 °C	55 °C	50–60 °C
Optimal pH	8.2		8.0–9.0	9.0	8.0	8.2–9.2
Temperature stability			<60 °C	<40 °C		<50 °C
pH Stability			6.0–7.0	5.5–8.5		around 6.5
Metal requirements	Mg^{2+}, Mn^{2+}	Zn^{2+}, Co^{2+}	Fe^{2+}, Co^{2+}?	Fe^{2+}		none
Induction			Hydantoin		5-Cyanoethylene-hydantoin	Hydantoin, dihydrouracil, various D,L-5-substituted hydantoins
Inhibition:						
EDTA		—	—	+		—
8-OH-Quinoline			+			—
o-Phenanthroline		+	+	+		—
p-OH-Mercuribenzoate			+	+		+
2,6-Dipicolinic acid		+		+		
2,2′-Dipyridyl			+	+		

acetone precipitations combined with an acid denaturation of the other proteins. Purification of microbial *N*-carbamoyl-D-amino acid amidohydrolases was carried out by several groups [31, 57, 58, 62]. The microorganisms used and the purification schemes are listed in Table 6.

4.3.2 Characterization

The characteristics of the enzyme extracts are summarized in Table 7. The substrate specificities are listed in Table 8.

Table 5. Substrate specificities of the D-hydantoinases

Reference	[69]	[68]	[75]	[9]	[77]	[57]
Hydantoin	+	16	2	13	5	5
5-Methylhydantoin		32(DL)	16	45		2(L)
1-Methylhydantoin			4			
5-*i*-Propylhydantoin		12(DL)		15	23	40(D)
5-*i*-Butylhydantoin		84(DL)		48		8(DL)
5-(2-Methylthioethylene)-hydantoin		20(DL)		48	41	100(D), 54(DL), 30(L)
5-(4-Aminobutylene)hydantoin						28(L)
5-Hydroxymethylenehydantoin						3(L)
5-Methoxymethylenehydantoin						30(DL)
5-Carboxymethylenehydantoin		0(DL)				5(L)
5-Carboxyethylenehydantoin		0(DL)				5(L)
5-(Carboxymethylidine)-hydantoin	—					
5,5-Dimethylhydantoin				0		
5-Phenylhydantoin		100(DL)		25		34(DL)
5-*p*-Hydroxyphenylhydantoin		3(DL)		16		33(D), 23(DL)
5-*m*-Hydroxyphenylhydantoin				5		19(DL)
5-*o*-Hydroxyphenylhydantoin						9(DL)
5-*p*-Chlorophenylhydantoin				19		
5-*m*-Chlorophenylhydantoin				10		
5-(2,4-Dichlorophenyl)-hydantoin				7		
5-*p*-Methoxyphenylhydantoin				4		
5-Benzylhydantoin		3(DL)				57(D)
5-*p*-Hydroxybenzylhydantoin					2	19(D)
5-(3,4-Dihydroxybenzyl)-hydantoin						18(DL)
5-(3,4-Dimethoxybenzyl)-hydantoin						8(DL)
5-Benzyloxymethylenehydantoin						14(DL)
5-(3,4-Methylenedioxybenzyl)-hydantoin						10(DL)
5,5-Diphenylhydantoin	—			0		
5-Indolylmethylenehydantoin		0(DL)				6(D)

Table 5. Continued

Reference	[69]	[68]	[75]	[9]	[77]	[57]
Dihydrouracil	+		10	100	100	9
5-Bromodihydrouracil			100			
5-Iododihydrouracil			74			
Dihydrothymine	+		16			
Barbituric acid	—					
Orotic acid	—					
Urocanic acid	—					
Imidazole 4,5-dicarboxylic acid	—					
4,5-Aminoimidazolecarboximide	—					
2-Thiohydantoin		0				
N-Carbamoyl-β-alanine	+		5			
N-Carbamoyl-β-alanine amide	—					
N-Carbamoyl-β-aminoisobutyrate	+		7			

+: conversion of the particular substrate; —: no conversion. If determined, the relative rates of conversion are specified. However, the relations are valid exclusively within one column. The configuration of the substrates is put in parentheses.

The D-specific *N*-carbamoylamino acid amidohydrolases from rat liver and *Clostridium uracilicum* are involved in the metabolic degradation of pyrimidines, while the enzyme from *Pseudomonas putida 77* is integrated into creatinine degradation pathway (see Sect. 3.3.4). In contrast to the D-hydantoinases, there are great differences concerning substrate specificities and induction. Furthermore this group of enzymes is cofactor-independent.

Table 6. Isolation of bacterial *N*-carbamoyl-D-amino acid hydrolases

Source	*Clostridium uracilicum*	*Agrobacterium radiobacter*	*Pseudomonas* sp. AJ-11220	*Pseudomonas putida 77*
Reference	[58]	[31]	[57]	[62]
Purification steps				
1st step	$MnCl_2$-precipitation		DEAE-Toyopearl	Ammonium sulfate
2nd step	Ammonium sulfate			DEAE-cellulose
3rd step	Acetone			1st crystallization
4th step	Ammonium sulfate			2nd crystallization
5th step	Acetone			
6th step	Hydroxylapatite			
Yield	18%		36%	63.2%
Purification factor	98.7		17	27.4
Remarks	Partially purified	Crude extract of disrupted cells	Partially purified	The homogeneous enzyme could be crystallized

Table 7. Properties of the *N*-carbamoyl-D-amino acid amidohydrolases

Source	Rat liver	*Clostridium uracilicum*	*Agrobacterium radiobacter*	*Pseudomonas* sp. AJ-11220	*Pseudomonas putida* 77
Reference	[70]	[58]	[31]	[57]	[62]
Molecular weight					102,000 Da
Subunits:					
Molecular weight					27,000 Da
Optimal temperature		30–35 °C	60 °C	55 °C	37 °C
Optimal pH	6.75–7.0	7.4–7.8	7.0	7.0	7–8
Temperature stability		<45 °C	<40 °C		<40 °C
pH Stability			7–9		6–7
Metal requirements	none	none	none		
Induction:		*N*-Carbamoyl-β-alanine	*N*-Carbamoyl-D-phenylglycine	5-Cyanoethylene-hydantoin*	1-Methylhydantoin*
Inhibition:					
EDTA		—	—		—
8-OH-Quinoline					—
o-Phenanthroline					—
2,2′-Dipyridil					—
N-Ethylmaleimid		—			
p-OH-Mercuribenzoate		—	+		+
$AgNO_3$					+
$HgCl_2$					+

* A sequential induction cannot be excluded

Table 8. Substrate specificities of the *N*-carbamoyl-D-amino acid amidohydrolases

Source Reference	Rat liver [70]	*Clostridium uracilicum* [58]	*Agrobacterium radiobacter* [31]	*Pseudomonas* sp. AJ-11220 [57]	*Pseudomonas putida* 77 [62]
N-Carbamoyl-β-alanine	+	+		8	
N-Carbamoyl-α-amino-*n*-butyric acid					0.9(DL)
N-Carbamoyl-α-aminoisobutyric acid					—
N-Carbamoyl-β-aminoisobutyric acid	+	—			
N-Carbamoylalanine	—(L)	—	63(D)	56(D)	7.3(DL)
N-Methyl-*N*-carbamoylalanine					12.5(DL)
N-Carbamoyl-*g*-aminoisobutyric acid	—				
N-Carbamoylglycine	—	—		8	9.8
N-Carbamoylsarcosine					100
N-Carbamoylvaline			42(D)	18(D)	0.7(D), +(DL), —(L)
N-Carbamoylleucine				61(D)	0.4(D)
N-Carbamoylisoleucine					—(L)
N-Carbamoylnorleucine					+(DL)
N-Carbamoylserine					—(L)
N-Carbamoylthreonine			0(L)		0.5(DL)
N-Carbamoylmethionine				100(D)	—(DL)
N-Carbamoylglutamic acid	—(L)		19(D), 0(L)	32(D)	—(L)
N-Carbamoylaspartic acid	—(L)	—			—(DL)
N-Carbamoylasparagine				48(D)	—(L)
N-Carbamoylproline	—(L)				
N-Carbamoylphenylalanine			62(D)	32(D)	1.4(DL)
N-Carbamoyl-3,4-dihydroxy-phenylalanine				62(D)	
N-Carbamoylphenylglycine			100(D), 0(L)	94(DL)	0.7(D)
N-Carbamoyl-*p*-hydroxyphenylglycine			98(D)	88(DL)	0.8(DL), +(D)
N-Carbamoyl-*p*-chlorophenyl-glycine			78(D)		

N-Carbamoyl-*p*-methoxyphenyl-glycine		62(D)		
N-Carbamoyl-*p*-methylphenyl-glycine				0.7(DL)
N-Carbamoyltryptophan			31(D)	1.7(DL)
N-Carbamoyltyrosine			30(D)	—(L)
N-Carbamoylhistidine				—(L)
N-Carbamoylthienylglycine		54(D)		
L-Citrulline	—			
N-Acetylalanine		0(D)		
N-Acetylphenylalanine		0(D)		

+ : conversion of the particular substrate; — : no conversion. If determined, the relative rates of conversion are specified. However, the relations are exclusively valid within the column. The configuration of the substrates is put in parentheses

5 L-Selective Hydrolysis of Hydantoin Derivatives

5.1 Microbial Conversion

5.1.1 History

Liebermann and Kornberg [45, 46] were the first to describe the L-specific microbial hydrolysis of a 5-substituted hydantoin derivative: L-aspartate was formed from L-5-carboxymethylenehydantoin by a *Clostridium oroticum* sp. Hassall and Greenberg [29] followed with a report on the L-specific enzymatic hydrolysis of L-5-(2′-carboxyethylene)hydantoin through the action of some unspecific Gram-negative coccoidal bacterium to form L-glutamic acid. This property was dependent on the presence of the substrate. Tsugawa et al. [36] found that the ability of enzymatic hydrolysis of 5-(2′-carboxyethylene)hydantoin was wide-spread in nature (see 3.2.2). The authors confirmed that the enzymes were inducible by the substrate. A new finding was that L-glutamic acid was formed from both, the L- and D-substituted hydantoin.

In 1964, a Japanese patent publication claimed the microbial L-specific hydrolysis of D,L-5-substituted hydantoins to a variety of L-amino acids, among them L-tryptophan, L-phenylalanine, L-leucine, L-isoleucine, and L-methionine [81]. Thereafter, Yamada et al. [49] described a *Bacillus coagulans* strain that yielded L-methionine from both enantiomers of 5-(2′-methylthioethylene)hydantoin; *N*-carbamoyl-D-methionine was found as a by-product.

5.1.2 Processes with Emphasis on the Formation of Aromatic L-Amino Acids

After these pioneering studies, further investigation has been focussed upon essential and economically attractive α-amino acids, like L-tryptophan and L-phenylalanine. All microorganisms discovered for this purpose show similar, but not identical characteristics with respect to inducibility of enzyme activities, substrate specificities, and reaction conditions.

In this group are enclosed bacterial genera of *Arthrobacter*, *Flavobacterium*, *Pseudomonas*, and *Bacillus* (see Sect. 5.1.3) with the property of L-specific conversion of 5-arylalkyl-substituted hydantoins (see also: Table 11). Both enantiomers are equally hydrolyzed because of spontaneous racemization of hydantoins under weakly alkaline conditions (see Sect. 2.3); the participation of a cell-bound hydantoin-racemase is, however, strongly suggested for several strains. For the formation of high enzyme activities, all species in question but *Bacillus* require 5-IMH as inducer along with traces of Mn^{2+} or Co^{2+} as cofactors. The reaction proceeds best at a slightly alkaline pH-value of 8.0 to 8.5 and often at elevated temperature of 37 °C to 50 °C. On the other hand, *Bacillus* is responsive to D,L-5-isopropylhydantoin as inducer, thus revealing diverging properties (see Sect. 5.1.3).

Based on early studies [25, 82–86] more detailed information was given by Yokozeki et al. [41, 42]. In shake-flask-experiments the L-tryptophan productivity from D,L-5-IMH was improved by use of mutants with the tryptophan degradation pathway genetically blocked (*Flavobacterium aminogenes* AJ-3940). The specific cell activity at 37 and 43 °C ranged from 0.4 to 0.6 mmol g^{-1} h^{-1} (L-Trp per CDM per time) [42].

In further studies [41], a complex medium containing corn steep liquor was used for cell growth; without supplementation of the cultivation medium with Co^{2+} or Mn^{2+}, 3.5 g l^{-1} D,L-5-IMH were necessary to obtain maximal activities at 30 °C within 16 h. Optimization tests on reaction conditions revealed an optimal pH and temperature of 8.5 and between 45 and 55 °C, resp.. Addition of Mn^{2+} or Co^{2+} to the transformation medium at above 1 mM led to increases in the yield of L-tryptophan by 10 to 20 %. This increment seems to be low. It has to be assumed, however, that the cells had already been almost completely induced, since they had been cultured in a complex medium which is likely to contain traces of these ions: for their presence during growth and enzyme induction, not during conversion, has proved to be vital to the function of the biocatalyst. But a stabilizing effect of these ions also under reaction conditions is conclusive (see below).

L-Tryptophan was found to inhibit its own formation at a concentration above 30 g l^{-1}. This could be prevented by the addition of inosine which forms a water-insoluble adduct with L-tryptophan. Other amino acids did not seem to inhibit the conversion of the corresponding hydantoin derivative. The yield was highest for the production of L-tryptophan, L-phenylalanine, L-3,4-dihydroxyphenylalanine, L-3,4-methylenedioxyphenylalanine under the conditions applied, whereas the conversion to L-5-hydroxytryptophan and L-3,4-dimethoxyphenylalanine was poor.

Studies carried out with another *Flavobacterium* sp. were published by Tanabe Seiyaku [87, 88]: L-Tryptophan and L-phenylalanine were produced from D- and L-5-IMH and D,L-5-benzylhydantoin by using cells or enzyme preparations at pH 8.5 and 37 °C. A detergent (Triton X-100, 0,02 %) was added to facilitate conversion. More details about this process can be drawn from [26]: High concentrations of the inducer, D,L-5-IMH (10 g l^{-1}), were applied to obtain high enzyme activities; the maximum was reached after an incubation of 30 to 40 h in shaken flasks in the presence of Mn^{2+} (30 mg l^{-1}). 1 g l^{-1} IMH exhibited virtually no inducing effect after that cultivation time. It was not examined whether at low concentrations the inducer might have been degraded long ago, followed by a loss of enzyme activities.

The reaction mixture contained 0.25 N NaCl, 4.7 g l^{-1} cells, 250 mM D,L-5-IMH (i.e. ca. 60 g l^{-1}!), and 0.26 mM cetyltrimethylammonium bromide, a cationic detergent, at pH 8.5 and 30 °C; Mn^{2+} was omitted. Remarkably, the initial specific cell activity, measured under standard conditions, was fairly high at about 11 mmol g^{-1} h^{-1} (L-Trp per CDM per time); a productivity of about 7 mmol g^{-1} h^{-1} was obtained by 90 % conversion within 6 h in shaken culture. These are presently the highest conversion rates documented for this reaction type. It has to be emphasized, though, that 0.26 mM cetyltrimethylammonium bromide was employed as a solubilizing and membrane active agent which may have increased the bioavailability of the substrate.

N-Carbamoyl-D-tryptophan was not hydrolyzed by the cells, although both, D- and L-IMH, as well as *N*-carbamoyl-L-tryptophan could serve as substrates. This implies that the D-carbamoyl-enantiomer must not be intermediarily formed from either enantiomer of the hydantoin. Unfortunately, chirality of the intermediates was not specified by the investigators. They argued that the first step in the production of L-tryptophan from D-5-IMH was racemization of D- to L-IMH.

The findings with *Flavobacterium* were in good agreement with the properties of several *Arthrobacter* sp., the first of which was introduced by Guivarch et al. [39]. After growth and induction of the cells of *Arthrobacter globiformis*, conversion of the hydantoin was optimal around pH 8.0 and 40 °C, whereas at pH 6.0 no reaction was observed. L-5-IMH and 5-benzylhydantoin were the most rapidly and equally well hydrolyzed substrates among the compounds tested. The specific cell activity was 1.08 mmol g^{-1} h^{-1} for L-Trp. L- and D-carbamoyl-amino acids were identified as the initial hydrolyzation products, the conversion of the D-enantiomer being the rate-limiting step for the total reaction. The authors postulated non-specific enzymatic hydrolysis and rapid enzymatic racemization of the hydantoins, but slow racemization on the level of the carbamoyl-amino acid.

Several patent applications [89–95] refer to the bioconversion of diverse D,L-5-arylated hydantoins by *Arthrobacter* sp., but also by *Flavobacterium* and *Pseudomonas* [91], resulting in the formation of L-tryptophan, L-phenylalanine, and L-tyrosine. Either D,L-*N*-carbamoyltryptophan or D,L-5-IMH were used as inducers at low concentrations (0.5–1.0 g l^{-1}) [89, 90, 92]. The inducing effect of the carbamoyl-intermediate may be explained by a hydantoinase-catalyzed reversed reaction to IMH which is then the active agent. In accordance with the findings of others, the conversion was Co^{2+}- or Mn^{2+}-dependent. The reaction was carried out in TRIS-buffer pH 7.0 and at 30 °C–35 °C.

Kato et al. [96] give also evidence of enzymatic racemization of 5-monosubstituted hydantoins by a hydantoin racemase of *Arthrobacter* sp. DK 200 (FERM P—7472), the enzyme was not isolated, however.

After early studies on *Arthrobacter aurescens* BH20 [6, 97, 98], a more efficient strain, *Arthrobacter* sp. DSM 3747, was isolated from soil [21]. A mutant with low L-tryptophan degrading, but high synthesizing activity has been the basis of the present development [40, 64, 65, 99–101].

The cells are grown at pH 7.0 and 30 °C in a mineral salts medium mainly containing glucose as C- and ammonium as N-source; a low quantity of yeast extract proved to be indispensable to obtain a high yield and concentration of biomass combined with high enzyme activities. In addition, the growth medium has to be supplemented with D,L-5-IMH and Mn^{2+}, but it was sufficient to add the inducer and Mn^{2+} not until the late growth phase. In consequence, retardation of growth caused by the inducing agents, the enrichment of degradation products from IMH, and a wasteful consumption of inducer could be overcome. A high cell concentration of 28–30 g l^{-1} was reached within 16 to 17 h [30, 102].

At optimal reaction conditions (phosphate buffer 0,05 M, pH 8.5 and 50 °C), the initial specific cell activity (mmol g^{-1} CDM per h) ranged from 1.4 (shaken flask) to 2.1 (bioreactor) for L-tryptophan; the specific cell productivity (mmol g^{-1} h^{-1}) by 90% conversion was 0.5 or 1.2, depending on the substrate concentration applied. The low value at a high initial substrate content (20 g l^{-1}) seems to be caused by product inhibition from L-tryptophan (see also [41]).

5.1.3 Other L-Stereoselective Processes Based on 5-Monosubstituted Hydantoin Derivatives

A strain of *Bacillus brevis* (AJ-12299] was found to convert several 5-alkylated hydantoin derivatives and *N*-carbamoyl-amino to the corresponding L-amino acids; both, the L- and D-enantiomer, were hydrolyzed L-specifically [103]. This microorganism was also reactive to 5-arylated hydantoins [38, 71]. Remarkably, the process is ATP-dependent, and both enzymes involved are L-specific. 5-Isopropylhydantoin served as the best inducer above other 5-alkylated hydantoins tested; 5-arylated derivatives were rather poor inducers.

Low concentrations of Mg^{2+}, Mn^{2+}, Fe^{2+} were found to activate the enzymes involved [71] (see Sects. 5.2.1 and 5.2.2). However, the conversion rates of all substrates tested were far below those usually found with *Flavobacterium* [41, 42] (see Sect. 5.1.2): at pH 7.0 and 30 °C the estimated cell activity was about 0.1 mmol g^{-1} h^{-1} for L-valine and L-leucine compared with $\leqq$0.6 mmol g^{-1} h^{-1} for L-tryptophan in *Flavobacterium*.

Although a mutant was genetically blocked with respect to the degradation of L-valine, it did not yield but less than 50% of L-valine from 10 g l^{-1} 5-isopropylhydantoin. The low reaction rate and incomplete conversion may be ascribed to the limited availability of ATP under the conditions of resting cells applied.

Recently a versatile *Nocardia* sp. (DSM 3306) able to L-specifically hydrolyze a large number of D,L-5-monosubstituted hydantoins was claimed [37]. The microorganism was isolated from soil with 5-isobutylhydantoin as sole *N*-source and appears to be most responsive to this substrate among all others rested to give L-leucine. Polar substituents on the 5-alkyl-residues were, however, not prohibitive, independent of whether they were positively or negatively charged. Remarkably, aryl-substituted hydantoins seem to be excluded from conversion to the corresponding L-amino acids. Neither inducing agents nor metal ions were specified. The specific productivities for L-leucine are poor under the conditions applied, ranging around 0.1 mmol g^{-1} h^{-1} in shake-flasks experiments.

5.1.4 L-Amino Acids from *N*-Carbamoyl-Compounds as Precursors

The methods to be described have to be differentiated from the categories outlined in Sects. 5.1.2 and 5.1.3, because enzymatic hydrolysis of 5-monosubstituted hydantoin derivatives has not been claimed by the investigators. On the other hand, all hydantoin-based processes can also be run with *N*-carbamoyl-amino acids as substrates unless their uptake into the cell is prohibited. Therefore, one can assume that the hydantoin structure is involved in the over-all reaction, since the L- as well as the D-carbamoyl-amino acid are often found to be transformed into the L-amino acid. This requires racemization which is likely to proceed on the hydantoin level, after cyclization of the substrate has taken place. Otherwise, racemization on the carbamoyl-stage would have to be postulated.

Microorganisms L-specifically catalyzing hydrolysis of *N*-carbamoyl-amino acids are *Pseudomonas* DK-910 [104], *Bacillus*, and *Vibrio* sp. [105], and *Alcaligenes faecalis* N-3972 [106]. A broad spectrum of unpolar and polar L-amino acids was prepared with *Pseudomonas*. *Bacillus* and *Vibrio* sp. were applied to produce L-methionine, and the *Alcaligenes* sp. to give L-phenylalanine. The formation of the enzyme activities of all strains was dependent on the presence of an appropriate carbamoyl-amino acid as inducer and Mn^{2+}.

Escherichia coli, transformed with a gene coding for *N*-carbamoyl-amino acid amidohydrolase, has acquired the ability to hydrolyze *N*-carbamoyl-amino acids. It was not specified whether both enantiomers were converted [146].

5.2 Isolation and Characterization of L-Hydantoinases

5.2.1 *Isolation*

Microbial L-hydantoinases were purified from *Flavobacterium* sp. AJ-3912 [63], *Bacillus brevis* AJ-12299 [71] and from *Arthrobacter* sp. DSM 3747 [63, 71, 74, 76]. Table 9 summarizes the particular enrichment and purification procedures.

5.2.2 Characterization

The three hydantoinases enriched seem to be different enzymes (see Sects. 3.2.2 and 3.3.4 for distribution and biological function). As shown in Table 10 the hydantoinase from *Bacillus brevis* AJ-12299 is ATP dependent. It also differs in its substrate specificity (see Table 11). While the L-hydantoinase from *Arthrobacter* sp. and *Flavobacterium* sp. mainly hydrolyze hydantoin derivatives containing aromatic side chains in positions 5, the hydantoinase from *Bacillus brevis* accepts a wide substrate spectrum including 5-substituted hydantoins with aliphatic residues (see also Sect. 5.1.3).

Table 9. Purification of bacterial L-hydantoinases

Source	*Flavobacterium* sp. AJ-3912	*Bacillus brevis* AJ-12299	*Arthrobacter* sp. DSM 3747
Reference	[63]	[71]	[76]
Purification steps:			
1st step	Protamine sulfate	DEAE-Toyopearl	Lq.-lq.-extraction
2nd step	Ammonium sulfate		Zetaprep 60 QAE
3rd step	DEAE-Toyopearl		Mono Q
4th step			Blue Sepharose
5th step			Superose 12
Yield	45%	75%	2.4%
Purification factor	50	14	253
Remarks	Partially purified	Partially purified	Purified to homogeneity

Table 10. Properties of bacterial L-hydantoinases

Source Reference	*Flavobacterium* sp. AJ-3912 [63]	*Bacillus brevis* AJ-12299 [71]	*Arthrobacter* sp. DSM 3747 [76]
Molecular weight			155–200 kDa
Subunits:			
Number			3 or 4
Molecular weight			55–62 kDa
Isoelectric pH			4.25–4.35
Optimal temperature	40 °C	50 °C	37 °C (30 °C)*
Optimal pH	9.7	8.0	8.8–9.0 (6.6–7.3)*
Metal requirements		Mg^{2+}, Mn^{2+} or K^+	Mn^{2+} or Co^{2+}
Other cofactors		ATP	
Induction			D,L-Indolyl-methylenehydantoin
Inhibition:			
EDTA			+
p-OH-Mercuribenzoate			+

* Ring closure reaction

Table 11. Substrate specificities of the L-hydantoinases

Source Reference	*Flavobacterium* sp. AJ-3912 [63]	*Bacillus brevis* sp. AJ-12299 [71]	*Arthrobacter* sp. DSM 3747 [74]
Hydantoin	<0.2		—
5-Methylhydantoin	0(DL)	57(L)	—(L)
1-Methylhydantoin			—
5-*i*-Propylhydantoin	<0.2(DL)	100(L), 100(DL), 75(D)	—(DL)
5-*i*-Butylhydantoin	<0.2(DL)	95(L)	+(DL)
5-*sec*-Butylhydantoin		111(L)	—(DL)
Proline derived hydantion			—(DL)
5-(2′-Methylthioethylene)-hydantoin	<0.2(DL)	130(L)	
5-(3′-Aminopropylene)-hydantoin			—(DL)
5-(4′-Aminobutylene)-hydantoin	0(DL)		—(DL)
5-Methoxymethylene-hydantoin	<0.2(DL)		
5-Carboxymethylene-hydantoin	0(DL)		—
5-Carboxymethylene-hydantoin	0(DL)		—(L)
5-Carbamoylethylene-hydantoin	<0.2(DL)		

Tables 11. Continued

Source Reference	*Flavobacterium* sp. AJ-3912 [63]	*Bacillus brevis* sp. AJ-12299 [73]	*Arthrobacter* sp. DSM 3747 [74]
5-Carbamoylmethylene-hydantoin	<0.2(DL)		—(DL)
5,5-Dimethylhydantoin			—
5,5-Dimethyl-1-hydroxy-methylhydantoin			—
5-Phenylhydantoin			—(DL)
5-*p*-Hydroxyphenylhydantoin			—(DL)
5-Benzylhydantoin	100(L), 0(D) 40(DL)	86(L)	+(DL)
5-*p*-Hydroxybenzylhydantoin	31(DL)	34(L)	+(DL)
5-(3,4-Dihydroxybenzyl)-hydantoin			+(DL)
5-(3,4-Dimethoxybenzyl)-hydantoin	0.7(DL)		
5-Benzyloxymethylene-hydantoin	2(DL)		+(DL)
5-(*S*-Benzylmercapto-methylene)hydantoin			+(DL)
5-(3′,4′-Methylenedioxy-benzyl)hydantoin	11(DL)		
5-(*p*-Hydroxymethylene-phenyl)-5-phenylhydantoin			—
5-Indolylmethylenehydantoin	48(DL)		+(DL)
5-(4-Imidazomethylene)-hydantoin	0(DL)		
5-(2-Thienyl)hydantoin			—(DL)
Inosine-5-monophosphate			—
Allantoin			—
Barbituric acid			—
Creatinine			—
Dihydroorotic acid			—
Oxindol			—
N-Carbamoylglycine			—
N-Carbamoyltryptophan			+
Tryptophanamid			—(L)
Tryptophanmethylester			—(L)
Stereo specificity:	Only stereoselective, relative rates relating to the L-configurated product	L-specific	

+: conversion of the particular substrate; —: no conversion. If determined the relative rates of conversion are specified. The relations are exclusively valid within one column. The configuration of the substrates is put in parentheses.

5.3 Isolation and Characterization of *N*-Carbamoyl-L-Amino Acid Amidohydrolases

5.3.1 Isolation

The first purification of an L-specific *N*-carbamoyl-amino acid amidohydrolase was carried out from *Zymobacterium oroticum* (= *Clostridium oroticum*) by Liebermann and Kornberg [46]. Further *N*-carbamoyl-L-amino acid amidohydrolase were purified from various microorganisms [57, 63, 71, 74]. Table 12 gives a survey on the bacteria used as sources and the purification procedures.

5.3.2 Characterization

In contrast to the *N*-carbamoyl-D-amino acid amidohydrolase, the L-specific group of enzymes is not characterized with respect to molecular weight, pH- and temperature stability. The properties obtainable from literature for partially purified preparations are summarized in Table 13.

With the exception of the ureidosuccinase [46] all *N*-carbamoyl-L-amino acid amidohydrolases show wide but differing substrate specificities (Table 14). There are also large differences between the relative rates of hydrolysis of the substrates listed. Therefore the five enzymes described seem to be different from each other.

6 Industrial Application

6.1 Manufacture and Use of α-Amino Acids — A Brief Survey

The industrial production of amino acids is estimated at about 560,000 t in 1987 [108]. L-α-amino acids account for the major part of this amount. Main products are (in t per year) monosodium-L-glutamate ca. 340,000, D,L-methionine ca. 140,000 and L-lysine 70,000. L-Glutamate and L-lysine are manufactured biologically. L-Glutamate is mainly used as a flavour enhancer, L-lysine and D,L-methionine are feed additives.

D,L-methionine is produced e.g. by alkaline hydrolysis of 5-(2′-methylthioethylene)hydantoin and used without racemate separation [3]. Another hydantoin derivative, 5-benzylidenehydantoin, is hydrolyzed to yield phenylpyruvate which is enzymatically transaminated to L-phenylalanine [109].

Amino acid production and application have been reviewed in several recent publications and books [1–3, 110, 111].

For comparison with hydantoin-based processes, current methods known for the production of amino acids are summarized in Table 15. Among the enzymatic processes the stereoselective application of hydrolases dominates. In the case of acylase, amidase, esterase and aminopeptidase, one enantiomer is selectively hydrolyzed whereas the other has to be recovered, racemized and again submitted to hydrolysis.

Table 12. Isolation of bacterial *N*-carbamoyl-L-amino acid amidohydrolases

Source	*Zymobacterium oroticum*	*Flavobacterium* sp. AJ-3912	*Pseudomonas* sp. AJ-11220	*Arthrobacter* sp. DSM 3747	*Bacillus brevis* AJ-12299
Reference	[46]	[63]	[57] (see 4.1.3)	[74]	[71]
Purification steps					
1st step	Protamine sulfate	Protamine sulfate		Lq.-lq.-extraction	DEAE-Toyopearl
2nd step	Ammonium sulfate	Ammonium sulfate		Zetaprep 60 QAE	
3rd step	Isoelectric fraction	DEAE-Toyopearl		Mono Q	
4th step				Superose 12	
Yield	65%	28%			77%
Purification factor	10.6	37			4.6
Remarks	Partially purified	Partially purified	Crude extract, no activity after DEAE-Toyopearl	Partially purified	Partially purified

Table 13. Properties of bacterial *N*-carbamoyl-L-amino acid amidohydrolases

Source	*Zymobacterium oroticum*	*Flavobacterium* sp. AJ-3912	*Pseudomonas* sp. AJ-11220	*Arthrobacter* sp. DSM 3747	*Bacillus brevis* AJ-12299
Reference	[46]	[63]	[57]	[74]	[71]
Optimal temperature		40 °C		45 °C	50 °C
Optimal pH	7.8–8.5	8.0		9.0	7.5
Metal requirements	Mn^{2+} or Fe^{2+}				Mn^{2+} or Fe^{2+}
Induction:		5-Indolylmethylene-hydantoin*	5-Cyanoethylene-hydantoin*	5-Indolylmethylene-hydantoin*	

* A sequential induction cannot be excluded

Stereospecific deacylation of *N*-acylamino acids is the only enzymatic process used on an industrial scale for the manufacture of optically pure amino acids, e.g. L-methionine, L-phenylalanine, L-valine, and D-valine [112–114].

The lyase processes are specific for one single product, except tryptophan synthetase and tyrosinase, which catalyze so-called β-replacement reactions [7, 115].

Amination reactions are carried out reductively either by using NADH-dependent dehydrogenases [111, 116] or pyridoxal phosphate dependent transaminases [117, 118]. The problem of NADH regeneration was elegantly solved by using PEG-bound NADH and formate dehydrogenase in an enzyme membrane reactor [116]. Generally the amino acid dehydrogenases have a high substrate specificity. Recently, however, a phenylalanine dehydrogenase with lower specificity has been reported [119, 120]. Moreover, dehydrogenases for the synthesis of D-α-amino acids from α-keto carboxylic-acids have also been described [121, 122].

Based on the hydrolysis of 5-monosubstituted hydantoins, three different procedures have been developed. The general advantages adhering to hydantoin processes are:

a. precursor (hydantoin derivatives) chemically accessible
b. low substrate specificity of the biocatalysts employed
c. high stereoselectivity for either D- or L-enantiomer of an α-amino acid
d. spontaneous racemization of the D,L-5-monosubstituted hydantoin derivatives under reaction conditions of enzymatic hydrolysis, therefore 100% yield of one enantiomer

To our knowledge, stereoselective hydrolysis of 5-monosubstituted hydantoins is industrially applied only to manufacture of D-*p*-hydroxyphenylglycine.

6.2 D-*p*-Hydroxyphenylglycine

Since the discovery of penicillin by Fleming in 1928, many analogues of this antibiotic had to be developed to overcome ever new bacterial resistances. An example for semisynthetic penicillins is amoxycillin, which carries D-*p*-hydroxyphenylglycine as a substituent (Fig. 6). The principle of semisynthesis has also been applied to the cephalosporines; both, cefadroxil and cefatrizine, comprise D-*p*-hydroxyphenylglycine.

The market size of these antibiotics is approximately three billion $ per year worldwide. This explains why several companies have developed enzymatic production processes for D-*p*-hydroxyphenylglycine which before has been produced exclusively by chemical synthesis and diastereomeric salt crystallization. Bayer AG employs immobilized subtilisin (EC 3.4.4.16) for racemate resolution of D,L-2-acetamido-4-hydroxyphenylacetic acid methylester [134]. DSM took another racemate separation process to the pilot stage which uses amidopeptidase and D,L-hydroxyphenylglycine amide [124].

Table 14. Substrate specificities of bacterial *N*-carbamoyl-L-amino acid amidohydrolases

Source Reference	*Zymobacterium oroticum* [46]	*Flavobacterium* sp. AJ-3912 [63]	*Pseudomonas* sp. AJ-11220 [57]	*Arthrobacter* sp. DSM 3747 [74]	*Bacillus brevis* sp. AJ-12299 [71]
N-Carbamoyl-β-alanine		3			
N-Carbamoyllysine		0			
N-Carbamoylalanine		0.5(L)	44(DL)	—	48(L)
N-Methyl-*N*-carbamoylalanine					
N-Carbamoyl-α-amino-isobutyric acid					
N-Carbamoylglycine		5(L)	0		
N-Carbamoylsarcosine					
N-Carbamoylvaline		2(L)	100(L)		100(L)
N-Carbamoylleucine		3(L)	98(L)		102(L)
N-Carbamoylisoleucine		2(L)			84(L)
N-Carbamoylserine		5(L)		—	
N-Carbamoylmethionine		24(L)	47(L)	—	73(L)
N-Carbamoylglutamic acid	—	0	3(L)	—	
N-Carbamoylaspartic acid	+(L), —(D)	0			
N-Carbamoylasparagine		1(L)	2(L)		
N-Carbamoylglutamine		1(L)			
N-Carbamoylproline		0			
N-Carbamoylphenylalanine		82(L)	10(L)	+	86(L)
N-Carbamoyl-3,4-dihydroxy-phenylalanine			0		
N-Carbamoyl-3,4-dimethoxy-phenylalanine		24(L)			+
N-Carbamoyl-3,4-methylene-dioxyphenylalanine		100(L)			+
N-Carbamoylphenylglycine				—	
N-Carbamoyl-*p*-chloro-phenylglycine					
N-Carbamoyltryptophan		55(L)	0	+	

N-Carbamoyltyrosine		59(L)	9(L)	+	45(L)
N-Carbamoyl-*O*-benzylserine		15(L)			
N-Carbamoyl-*O*-methylserine		13(L)			
N-Carbamoylhistidine		0			
L-Citrulline	—				
N-Acetyltryptophan				—	
Glycyltryptophan				—	
Tryptophanamid				—	
Tryptophanmethylester				—	
N-Formyltryptophan				+	
N-Formylphenylalanine				+	
N-Formylmethionine				—	

+ : conversion of the particular substrate; — : no conversion. If determined, the relative rates of conversion are specified. The relations are exclusively valid within one column. The configuration of the substrates is put in parentheses.

Table 15. Processes for manufacture of optically pure α-amino acids

Process	Substrate	Enantiomer D-	Enantiomer L-	Examples	References*
Batch growth	Cheap C- and N-sources	—	+	L-Glutamate, L-lysine from melasse	[3]
Batch growth	Precursor	—	+	L-Tryptophan from anthranilic acid	[123]
Extraction	Protein hydrolysates lysates	—	+	L-Cysteine and L-tyrosine from hair	[3]
Racemate separation					
• fractionated crystallization	Diastereomeric salts of D,L-Amino acids	+	+	D-Phenylglycine	[124]
• acylase	D,L-*N*-Acetyl amino acids	+	+	L-Methionine	
• esterase	D,L-Amino acid esters	+	+	D-Tryptophan	[116]
• aminopeptidase	D,L-*N*-Acetamido acid esters	+	+	D-Valine	
• amidase	D,L-Amino acid amides	+	+	D,L-Tryptophan	[125]
Hydantoin hydrolysis					
• hydantoinase	D,L-5-Monosubstituted hydantoins	+	+	*N*-Carbamoyl-D-valine	[126]
				N-Carbamoyl-D-*p*-hydroxy-phenylglycine	[127]
• hydantoinase	dito	+	+	L-Tryptophan	[6]
+ carbamoylase				D-Valine	[128]
— carbamoylase	D- or L-*N*-Carbamoyl-amino acid	+	+	D-Valine	[31]
Asymmetric synthesis	Unsaturated precursor	+	+	L-DOPA using Wilkinson-catalysts	[123]
	Lactim ether	+	+	(Schöllkopf)	
Lyases					
• aspartase	Fumaric acid	—	+	L-Aspartate	[113]
• tryptophan synthetase	L-Serine, indole	—	+	L-Tryptophan	[129]
	L-Serine + selenol	—	+	L-Selenocysteines	[115]
• tryptophanase	Pyruvate, indole	—	+	L-Tryptophan	[130]
• tyrosinase	Pyruvate, phenol	—	+	L-Tyrosine	[131]

Reductive amination (NADH-dependent)					
• dehydrogenases	α-Keto acids	+	+	L-Phenylalanine L-*Tert*-leucine	[118] [119, 120]
Transamination (PLP-dependent)					
• transaminases	α-Keto acids	?	+	L-Phenylalanine D-*p*-Hydroxyphenylglycine	[132] [133]

* References given in this table are meant as examples and by no means complete

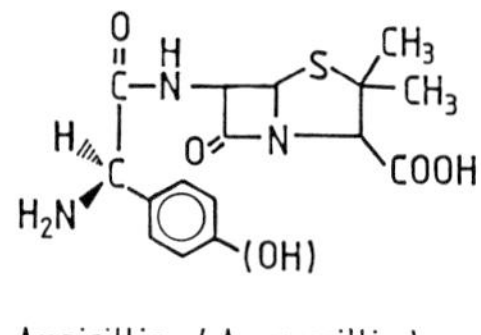

Fig. 6. Structural formulae of the semisynthetic antibiotics ampicillin, and amoxicillin

Yamada together with Kanegafuchi Chemical Industries [8, 135] (Fig. 7a) and Cecere et al. at Snamprogetti [67, 136, 137] (Fig. 7b) pioneered the field of industrial application of enzymatic D-specific hydantoin hydrolysis (see also Sect. 3.3.3). Kanegafuchi started commercial production of D-*p*-hydroxyphenylglycine at a Singapore plant in 1983. Another hydantoin process for this compound was developed at Ajinomoto Co. [128] (Fig. 7c).

Tramper and Luyben [138] compared the feasibility of the five processes outlined above. Because of the higher number of reaction steps, they excluded the Bayer and DSM processes from further evaluation, although the Bayer process has the advantage of high substrate concentration and thus a smaller bioreactor. Among the three hydantoin processes (Fig. 7), they selected the Kanegafuchi process as the most feasible because of the lower stability and higher cost of the calf liver D-hydantoinase (Snamprogetti) and the low enzyme activity of the Ajinomoto biocatalyst.

The Kanegafuchi process was then evaluated in detail by design of a process scheme and preparation of a cost analysis [139]. The following assumptions were made: annual (300 workdays) production capacity of 300 t, overall yield of 90%, bioconversion productivity of 5.4×10^{-2} mol kg^{-1} h^{-1} immobilized bio-

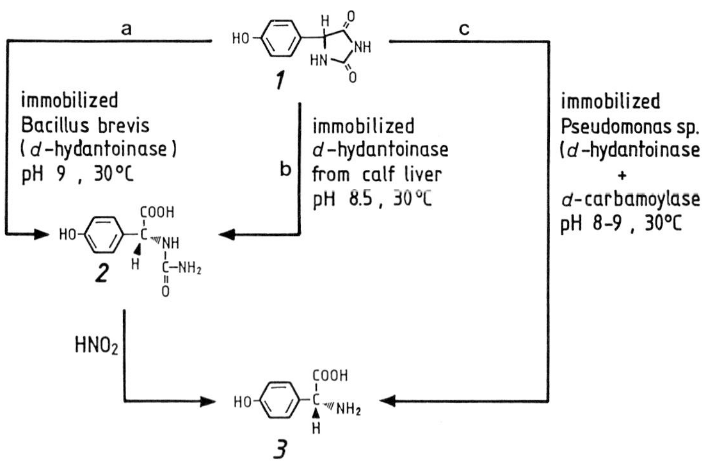

Fig. 7. Enzymatic hydrolysis of D,L-5-(4′-hydroxyphenyl)hydantoin (**1**) to *N*-carbamoyl-D-*p*-hydroxyphenylglycine (**2**) and D-*p*-hydroxyphenylglycine (**3**). **a.** Kanegafuchi process, **b.** Snamprogetti process, **c.** Ajinomoto process. (References see text)

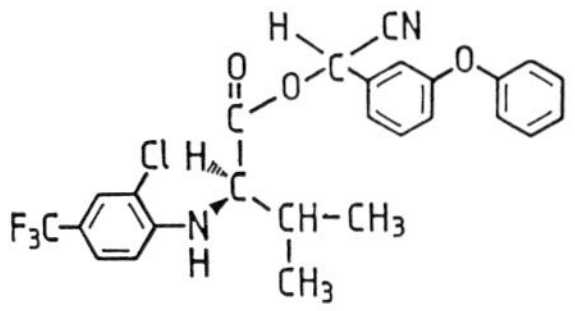

Fig. 8. Structural formula of the contact insecticide fluvalinate

catalyst (= 8.95 g kg^{-1} h^{-1} biocatalyst for *p*-hydroxyphenylglycine). Characteristics of the process were assumed to be: a. continuous production of biomass, b. immobilization in Ca-alginate, c. fed-batch bioconversion of hydantoin and d. batch hydrolysis of *N*-carbamoyl-D-*p*-hydroxyphenylglycine. In conclusion, these authors estimate a final product price of 60–80 Fl kg^{-1}, assuming precursor costs of 30 Fl kg^{-1}. Therefore, under the assumptions made, biotechnology is competitive with the chemical processes currently used.

The establishment of large scale production of D-phenylglycine and D-valine by the same process has been reported [140]. D-valine is used as a precursor for the insecticide fluvalinate (Fig. 8).

6.3 Prospects for Application to Other Amino Acids

Enzymatic hydantoin hydrolysis shows most of the characteristics important to the industrialization of a biochemical process. The precursors, hydantoin derivatives, are easily synthesized by well known procedures. In the case of both, D- and L-selective hydrolysis, the actions of the respective enzymes are highly stereoselective, but with low substrate specificity. Racemization of the D,L-5-substituted hydantoins occurs spontaneously or enzymatically, thus fugitive substituents are retained. Enzymatic or microbial hydrolysis of D,L-substituted hydantoins to either D- or L-amino acids of high enantiomerical purity is a

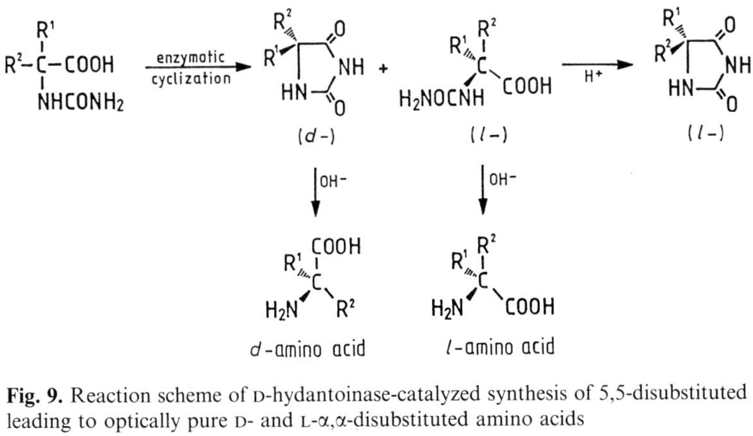

Fig. 9. Reaction scheme of D-hydantoinase-catalyzed synthesis of 5,5-disubstituted D-hydantoins leading to optically pure D- and L-α,α-disubstituted amino acids

versatile approach of great promise of commercialization. Particularly, speciality amino acids with low market volume may be rendered accessible by this method.

The industrial interest in enzymatic D-specific hydantoin hydrolysis for the production of a variety of D-α-amino acids is reflected by the patent literature [127, 130, 141, 143]. It is not known, however, whether these results have already found industrial application.

For L-selective hydantoin cleavage, there also exist several references in the patent literature (see Sect. 5.1.2). To our knowledge, none of these results have yet been taken to the industrial scale.

The productivity of the process for the production of L-tryptophan from D,L-5-indolylmethylene hydantoin developed in our group (see Sect. 5.1.2) has been continuously increased. Higher activities have been reported by [26]. Taking into account these productivities, the costs for the hydantoin process may approach the current market price of L-tryptophan.

To conclude with, an interesting variant of the use of D-hydantoinase has been described by Watanabe et al. [144]. They use the enzyme for catalysis of the reverse reaction, i.e. D-selective synthesis of the hydantoin from the racemic mixture of an α,α-disubstituted *N*-carbamoylamino acid (Fig. 9). α,α-Disubstituted amino acids, e.g. L-α-methyl-DOPA, act as antagonists of the corresponding α-H-amino acids [110]. The interest in these compounds is increasing.

In a similar way, D-hydantoinase was exploited to obtain diverse L-amino acids [145]: From a mixture of D- and L-*N*-carbamoylamino acids, the D-enantiomer was enzymatically transformed into the D-hydantoin, while the L-chiral compound was chemically hydrolyzed; thereafter, the D-hydantoin derivative can chemically be cleaved to give a racemic mixture of carbamoyl-amino acid.

6.4 Diagnostic Uses

A 1-methylhydantoinase can be used in a coupled enzyme test for the quantitative determination of creatinine in blood samples via creatinine deaminase catalyzed formation of 1-methylhydantoin (Fig. 10). The sarcosine finally formed in this reaction sequence is determined photometrically, following NAD-dependent dehydrogenation [13].

7 Conclusion

Enzymatic hydantoin cleavage either by intact cells or by isolated enzymes is already important in the production of D-amino acids and will extend its importance to L-amino acids. The wide applicability to a broad substrate range together with their high enantioselectivity and the complete conversion of 100% of a hydantoin racemate into the optically pure enantiomer render these processes very attractive. Immobilization of cells or enzymes may be applicable as already

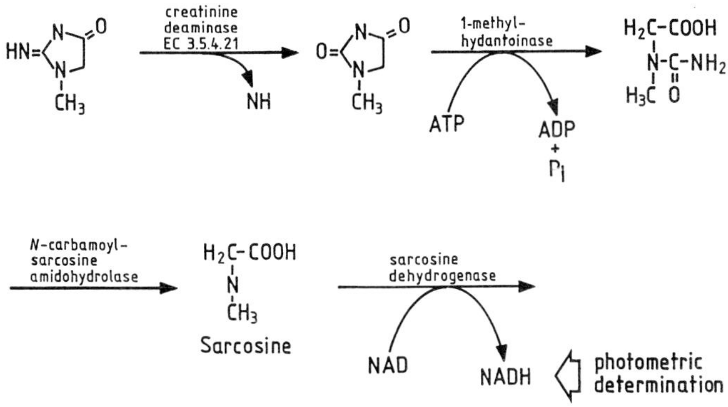

Fig. 10. Reaction scheme for enzymatic determination of creatinine by use of 1-methylhydantoinase

demonstrated for D-*p*-hydroxyphenylglycine and have to be developed for other cases.

The mechanism of enzymatic racemization is an interesting problem still to be investigated. Information on natural function and on genetics is lacking for L-selective hydantoinases and *N*-carbamoyl-L-amino acid amidohydrolases.

8 Acknowledgements

Part of the work of the authors has been supported by a grant of Bundesministerium für Forschung und Technologie (Forschungsvorhaben 03 87 35). The substrates used in this work were synthesized by Prof. Dr. K. Krohn, Technische Universität Braunschweig.

9 Symbols and Explanations

ATP	adenosine triphosphate
C-	carbon-
CDM	cell dry mass
IMH	5-indolylmethylenehydantoin
N-	nitrogen
NAD/NADH	nicotine amide adenine dinucleotide
PLP	pyridoxal phosphate
Trp	tryptophan

10 References

1. Aida K, Chibata I, Nakayama K, Takinami K, Yamada H (1986) Biotechnology of amino acid production, Prog. Ind. Microbiol., vol 24, Elsevier, Amsterdam
2. Kleemann A, Leuchtenberger W, Hoppe B, Tanner H (1985) In: Ullmann's Encyclopedia of Industrial Chemistry, vol 2, Verlag Chemie, Weinheim, p 57
3. Hoppe B, Martens J (1983) Chemie in unserer Zeit 17: 41
4. Jakubke HD, Jeschkeit H (1982) Aminosäuren, Peptide, Proteine. Verlag Chemie, Weinheim
5. Kleemann A (1982) Chem. Ztg. 106: 151
6. Syldatk C, Cotoras C, Möller A, Wagner F (1986) BTF-Biotechforum 3: 9
7. Yamada H, Shimizu S (1988) Angew. Chem. 100: 640
8. Yamada H, Takahashi S, Kii Y, Kumagai H (1978) J. Ferment. Technol. 56: 484
9. Takahashi S, Kii Y, Kumagai H, Yamada H (1978) Agric. Biol. Chem. 56: 492
10. Takahashi S, Ohashi T, Kii Y, Kumagai H, Yamada H (1979) J. Ferment. Technol. 57: 328
11. Yamada H, Shimizu S, Shimada H, Yoshiki T, Tani Y, Takahashi S, Ohashi T (1980) Biochimie 62: 395
12. Shimizu S, Shimada H, Takahashi S, Ohashi T, Tani Y, Yamada H (1980) Agric. Biol. Chem. 44: 2233
13. Siedel J, Deeg R, Röder A, Zieghorn J, Möllering H, Gauhl H (1985) Germ. Pat. DE 3406770 A1
14. Baeyer A (1861) Ann. 117: 178
15. Ajinomoto Co (1979) Jap. Pat. 80 136279
16. Kleemann A, Samson M (1980) Germ. Pat. DE 3043250 A1
17. Kleemann A, Samson M (1980) Germ. Pat. DE 3043259 A1
18. Drauz K, Kleemann A, Samson M (1984) Chem. Ztg. 12: 391
19. Ware E (1950) Chem. Rev. 46: 403
20. Morin A, Hummel W, Kula MR (1986) Biotechnol. Lett. 8: 573
21. Groß C, Syldatk C, Wagner F (1987) Biotechnol. Tech. 1: 85
22. Wanru S (1983) Acta Microbiologica Sinica 23: 133
23. Yokozeki K, Nakamori S, Eguchi C, Yamada K, Mitsugi K (1987) Agric. Biol. Chem. 51: 355
24. Yokozeki K, Nakamori S, Yamanaha S, Eguchi C, Mitsugi K, Yoshinaga F (1987) Agric. Biol. Chem. 51: 715
25. Sano K, Yokozeki K, Eguchi C, Kagawa T, Noda I, Mitsugi K (1977) Agric. Biol. Chem. 41: 819
26. Nishida Y, Nakamichi K, Nabe K, Tosa T (1987) Enzyme Microb. Technol. 9: 721
27. Möller A, Syldatk C, Schulze M, Wagner F (1988) Enzyme Microb. Technol. 10: 618
28. Morin A, Hummel W, Kula MR (1987) J. Gen. Microbiol. 133: 1201
29. Hassall H, Greenberg DM (1963) J. Biol. Chem. 338: 3325
30. Syldatk C, Mackowiak V, Höke H, Groß C, Dombach G, Wagner F, J. Biotechnol. (in press)
31. Olivieri R, Fascetti E, Angelini L, Degen L (1979) Enzyme Microbiol. Technol. 1: 201
32. Sobotka H (1932) US Pat. 1861458
33. Bernheim F, Bernheim MLC (1946) J. Biol. Chem. 163: 683
34. Bernheim F (1947) Federat. Proc. 6: 238
35. Eadie GS, Bernheim F, Bernheim MLC (1949) J. Biol. Chem. 181: 449
36. Tsugawa R, Okumura S, Ito T, Katsuga N (1966) Agric. Biol. Chem. 30: 27
37. Klages U, Weber A, Wilschowitz L (1987) Germ. Pat. DE 3702384 A1
38. Yamashiro A, Yokozeki K, Kano H, Kubota K (1988) Agric. Biol. Chem. 52: 451
39. Guivarch M, Gillonier C, Brunie JC (1980) Bull. Soc. Chim. Fr. 1–2: 91
40. Syldatk C, Cotoras D, Dombach G, Groß C, Kallwaß H, Wagner F (1987) Biotechnol. Lett. 9: 25
41. Yokozeki K, Sano K, Eguchi C, Iwagami H, Mitsugi K (1987) Agric. Biol. Chem. 51: 729

42. Yokozeki K, Sano K, Eguchi C, Yamada K, Mitsugi K (1987) Agric. Biol. Chem. 51: 363
43. International Union of Biochemistry (ed.) Enzyme Nomenclature 1984, Academic, New York (1984)
44. Akamatsu M (1960) J. Biochem. 47: 809
45. Liebermann I, Kornberg A (1954) J. Biol. Chem. 207: 911
46. Liebermann I, Kornberg A (1955) J. Biol. Chem. 212: 909
47. Yates RA, Pardee AB (1956) J. Biol. Chem. 221: 743
48. Taylor H, Taylor ML, Balch WE, Gilchrist P (1976) J. Bacteriol. 127: 863
49. Yamada H, Oishi K, Aida K, Uemura T (1969) Nippon Nogein Kagaku Kaishi 43: 528
50. Vogels GP, Trijbels F, Uffkint A (1966) Biochim. Biophys. Acta 122: 482
51. Lee KW, Roush AH (1964) Arch. Biochem. Biophys. 108: 460
52. Okumura I, Kondo K, Miyake Y, Itaya K, Yamamoto T (1976) J. Biochem. 79: 1013
53. Trijbels F, Vogels GD (1966) Biochim. Biophys. Acta 113: 292
54. van der Trift L, Vogels GD, van der Trift C (1975) Biochim. Biophys. Acta 391: 240
55. Morin A, Hummel W, Kula MR (1986) Appl. Microbiol. Biotechnol. 25: 91
56. Olivieri R, Fascetti E, Angelini L, Degen L (1983) Biotechnol. Bioeng. 23: 2173
57. Yokozeki K, Kubota K (1987) Agric. Biol. Chem. 51: 721
58. Campbell LL (1960) J. Biol. Chem. 235: 2375
59. Wanru S (1983) Acta Microbiologica Sinica 23: 257
60. Hayaishi O, Kornberg A (1952) J. Biol. Chem. 197: 717
61. Yamada H, Shimizu S, Kim JM, Shinmen Y, Sakai T (1985) FEMS Microbiol. Lett. 30: 337
62. Kim JM, Shimizu S, Yamada H (1986) J. Biol. Chem. 261: 11832
63. Yokozeki K, Hirose Y, Kubota K (1987) Agric. Biol. Chem. 51: 737
64. Syldatk C, Müller, R, Dombach G (1988) In: Abstr. Book 8th Int. Biotechnol. Symp., Paris, 1988, p 154
65. Syldatk C, Mackowiak V, Dombach G, Groß C, Röhrmann A, Wagner F (1987) In: Neijssel OM, van der Meer RR, Luyben KCAM (eds) Proc. 4th Europ. Cong. Biotechnol., vol 2. Elsevier, Amsterdam, p 245
66. Vogels GD (1966) Biochim. Biophys. Acta 113: 277
67. Dinelli D, Marconi W, Cecere F, Galli G, Morisi F (1978) In: Pye EK, Weethall HH (eds). Enzyme Engineering 3, Plenum, New York, p 477
68. Cecere F, Galli G, Morisi F (1975) FEBS-Lett. 57: 192
69. Wallach DP, Grisolia S (1957) J. Biol. Chem. 226: 277
70. Caravaca J, Grisolia S (1958) J. Biol. Chem. 231: 357
71. Yamashiro A, Kubota K, Yokozeki K (1988) Agric. Biol. Chem. 52: 2857
72. Gaebler OH, Keltch AK (1926) J. Biol. Chem. 70: 763
73. Wada M (1934) Proc. Imp. Acad. Japan 10: 17
74. Syldatk C, Müller R, Wagner F In: Stoffumwandlungen mit Biokatalysatoren, BMFT-Statusreport 1988 (Bundesministerium für Forschung und Technik, ed.) (in press)
75. Brooks KP, Jones EA, Kim BD, Sander EG (1983) Arch. Biochem. Biophys. 226: 469
76. Dombach G (1989) Doctoral thesis, Technical University Braunschweig
77. Morin A, Hummel W, Schütte H, Kula MR (1986) Biotechnol. Appl. Biochem. 8: 564
78. Denki Kagaku (1986) Jap. Pat. 61 285996
79. Rijnierse VFM, van der Drift C, Vogels GD (1977) Can. J. Microbiol. 23: 633
80. Battilotti M, Barberini U (1988) J. Mol. Cat. 43: 343
81. Ajinomoto Co (1964) Jap. Pat. 67 013850
82. Sano K, Yokozeki K, Yamada K, Yasuda N, Eguchi C, Noda I, Mitsugi K (1977) Jap. Pat. 77 15891
83. Yokozeki K, Sano K, Mitsugi K, Yamada K, Kagawa T, Eguchi C, Noda I (1977) Jap. Pat. 77 07490
84. Yokozeki K, Sano K, Yamada K, Eguchi C, Togo K, Noda I, Mitsugi K, Tamura F (1977) Jap. Pat. 77 18887
85. Mitsugi K, Sano K, Yokozeki K, Yamada K, Noda I, Kagawa T, Eguchi C, Yasuda N, Tamura F, Togo K (1977) US Pat. 4016037
86. Ajinomoto Co (1977) French Pat. 2329644

87. Tanabe Seiyaku Co (1983) Jap. Pat. 59 156293
88. Tosa T, Nabe K, Nishida Y, Nakamichi K (1984) Jap. Pat. 61 12296
89. Miyoshi T, Kitagawa H, Kato M, Chiba S (1985) Europ. Pat. 0159866 A2
90. Kato M, Omine H, Yamamoto H, Miyoshi T (1985) Jap. Pat. 61 285995
91. Denki Kagaku (1985) Jap. Pat. 61 285996
92. Kato M, Omine H, Yamamoto H, Miyoshi T (1985) Jap. Pat. 61 285997
93. Akimoto T, Watanabe M, Nagasaki S, Hirata M (1985) Jap. Pat. 62 00270
94. Akimoto T, Watanabe M, Nagasaki S, Hirata M (1985) Jap. Pat. 62 00271
95. Akimoto T, Watanabe M, Nagasaki S, Hirata M (1985) Jap. Pat. 62 03792
96. Kato M, Kitagawa H, Miyoshi T (1985) Jap. Pat. 62 122591
97. Cotoras D, Wagner F (1984) In: Proc. 3rd Europ. Congr. Biotechnol. vol 1, Verlag Chemie, Weinheim, p 350
98. Cotoras D (1985) Doctoral Thesis, Technical University Braunschweig
99. Groß C (1987) Doctoral Thesis, Technical University Braunschweig
100. Groß C, Syldatk C, Wagner F (1987) In: Neijssel OM, van der Meer RR, Luyben KCAM (eds) Proc. 4th Europ. Congr. Biotechnol., vol. 2, Elsevier, Amsterdam, p 248
101. Höke H, Höltmann W, Wagner F, Cotoras D, Syldatk C, Groß C, Dombach G, Wagner T (1988) Germ. Pat. DE 3712539 C2
102. Höke H In: Stoffumwandlungen mit Biokatalysatoren, BMFT-Statusreport 1988 (Bundesministerium für Forschung und Technik, ed.), (in press)
103. Yokozeki K, Yamashiro A, Kubota K, Kano H (1988) Jap. Pat. 63 24895
104. Miyoshi T, Kitagawa H, Kato M, Chiba S (1986) Jap. Pat. 61 09292
105. Ishikawa T, Horikoshi K, Koyama Y, Kimura H (1987) Jap. Pat. 62 275696
106. Ishida H, Horikoshi T (1988) Jap. Pat. 63 63395
107. Usui N, Yokozeki K, Kawashima N, Ei H, Kubota K (1988) Jap. Pat. 63 24895
108. Leuchtenberger W, Plöcker U (1988) Chem. Ing. Tech. 60: 16
109. Calton GJ (1987) In: Neijssel OM, van der Meer RR, Luyben KCAM (eds) Proc. 4th Europ. Congr. Biotechnol., vol 4 Elsevier, Amsterdam, p 693
110. Kamphuis J, Klostermann M, Schoemaker HE, Boesten WHJ, Meijer EM (1987) In: Neijssel OM, van der Meer RR, Luyben KCAM (eds) Proc. 4th Europ. Congr. Biotechnol., vol 4 Elsevier, Amsterdam, p 331
111. Leuchtenberger W (1987) In: Neijssel OM, van der Meer RR, Luyben KCAM (eds) Proc. 4th Europ. Congr. Biotechnol., vol 4, Elsevier, Amsterdam, p 701
112. Leuchtenberger W (1984) Chemie, Labor, Betrieb 35: 278
113. Chibata I (1983) In: Hollaender A, Laskin AI, Rogers P (eds) Basic biology of new developments in biotechnology, Plenum, New York, p 465
114. Izumi Y, Chibata I, Itoh T (1978) Angew. Chem. 90: 187
115. Soda K, Tanaka H (1987) Biotech. 1: 93
116. Wandrey C (1987) In: Neijssel OM, van der Meer RR, Luyben KCAM (eds) Proc. 4th Europ. Congr. Biotechnol., vol 4, Elsevier, Amsterdam, p 171
117. Hsiao HY, Wei TT, Anderson DM (1986) Int. Pat. WO 8605515
118. Anderson DM (1987) Biotech. 1: 41
119. Asano Y, Nakazawa A (1987) Agric. Biol. Chem. 51: 2035
120. Asano A, Nakazawa A (1985) Agric. Biol. Chem. 49: 3631
121. Wandrey C, Wichmann R, Leuchtenberger W, Kula MR, Bückmann A (1981) Germ. Pat. DE 2930070 A1
122. Wandrey C, Wichmann R, Leuchtenberger W, Kula MR, Bückmann A (1981) Germ. Pat. DE 2930087 A1
123. Showa Denko (1984) Jap. Pat. 59 125892
124. Meijer EM, Boesten WHJ, Schoemaker HE, van Balken JAM (1985) In: Tramper J, van der Plas HC, Linko P (eds) Biocatalysts in organic syntheses, studies in organic chemistry 22, Elsevier, Amsterdam, p 135
125. Nakai M, Obshima TN, Kimura T, Omata T, Iwamoto N (1982) Europ. Pat. 0043211 A2 A3
126. Jacob E, Henco K, Marcinowski S, Schenk G (1987) Germ. Pat. DE 3535987

127. Lungershausen R, Martin C, Marcinowski S, Siegel H, Kuesters W (1982) Germ. Pat. DE 3031151 A1
128. Nakamori S, Yokozeki K, Mitsugi K, Eguchi C, Iwagami H (1980) US Pat. 4211840
129. Bang WG, Behrendt U, Lang S, Wagner F (1983) Biotechnol. Bioeng. 25: 1013
130. Watanabe W, Snell EE (1972) Proc. Nat. Acad. Sci. 69: 1086
131. Yamada H (1987) In: Neijssel OM, van der Meer RR, Luyben KCAM (eds) Proc. 4th Europ. Congr. Biotechnol., vol 4, Elsevier, Amsterdam, p 689
132. Rozzell D (1985) Europ. Pat. 0135846 A2
133. van den Tweel WJJ, Ogg RLHP, de Bont JAM (1987) Dutch Pat. 8702449
134. Schutt H, Schmidt-Kastner G, Arens A, Preiss M (1985) Biotechnol. Bioeng. 27: 420
135. Takahashi H, Takahashi S, Ohashi T, Yoneda K, Watanabe K (1987) Jap. Pat. 62 25990
136. Cecere F, Marconi W, Morisi F, Rappuoli B (1978) Germ. Pat. DE 2615594 A1
137. Degen L, Viglia A, Fascetti E, Perricone E (1976) Germ. Pat. DE 2631048
138. Tramper J, Luyben KCAM (1982) PT/Procestechniek 37: 97
139. Tramper J, Luyben KCAM (1984) PT/Procestechniek 39: 61
140. Yamada H, Shimizu S (1985) In: Tramper J, van der Plas HC, Linko P (eds) Biocatalysts in organic syntheses, studies in organic chemistry 22, Elsevier, Amsterdam, p 19
141. Gillonier C, Guivarch M (1980) Germ. Pat. DE 3018584 A1
142. Takeichi M, Hagiwara T, Tarukawa H, Tawaki S (1985) Jap. Pat. 61 212292
143. Tawaki S, Tarukawa H, Aikawa T, Takeichi M (1988) Jap. Pat. 63 112990
144. Kanegafuchi Kagaku Kogyo (1985) Europ. Pat. 0175312 A2
145. Olivieri R, Eletti BG, Fascetti E, Centini F (1988) Europ. Pat. 0152977
146. Ajinomoto Co (1986) Jap. Pat. 63 24894

Plastics from Bacteria and for Bacteria: Poly(β-Hydroxyalkanoates) as Natural, Biocompatible, and Biodegradable Polyesters

Helmut Brandl[1]*, Richard A. Gross[2 #], Robert W. Lenz[2], and R. Clinton Fuller[1]

[1] Department of Biochemistry, University of Massachusetts, Amherst, MA 01003, USA

[2] Department of Polymer Science and Engineering, University of Massachusetts, Amherst, MA 01003, USA

1 Microbial Formation of Poly(β-Hydroxyalkanoates) ... 78
1.1 Structure of PHA ... 78
1.2 Occurrence of PHA in Microorganisms ... 78
1.3 Environmental Conditions Affecting PHA Formation ... 80
1.4 PHA as Intracellular Inclusion Bodies ... 81
2 Characterization and Properties of PHA ... 82
2.1 Physical and Chemical Characterization ... 82
2.2 Analytical Methods ... 83
2.3 Functional PHA ... 84
3 PHA as Biodegradable Plastics ... 85
3.1 Industrial Production of PHA ... 85
3.2 Biodegradation of PHA ... 87
4 Biodegradable Plastics and Solid Waste Disposal ... 89
4.1 Ecological Aspects ... 89
4.2 Disposal of Plastics Waste ... 89
5 Acknowledgement ... 90
6 Abbreviations ... 90
7 References ... 91

A wide variety of different types of microorganisms are known to produce intracellular energy and carbon storage products which have been generally described as being poly(β-hydroxybutyrate), PHB, but which are, more often than not, copolymers containing different alkyl groups at the β-position. Hence, PHB belongs to the family of poly(β-hydroxyalkanoates), PHA, all of which are usually formed as intracellular inclusions under unbalanced growth conditions. Recently, it became of industrial interest to evaluate PHA polyesters as natural, biodegradable, and biocompatible plastics for a wide range of possible applications such as surgical sutures or packaging containers. For industrial applications, the controlled incorporation of repeating units with different chain lengths into a series of copolymers is desirable in order to produce polyesters with a range of material properties because physical and chemical characteristics depend strongly on the polymer composition. Such "tailor-made" copolymers can be produced under controlled growth conditions, in that if a defined mixture of substrates for a certain type of microorganisms is supplied, a well defined and reproducible copolymer is formed.

* Present address: Institute of Plant Biology, Zollikerstraße 107, 8008 Zürich, Switzerland

Present address: Department of Chemistry, University of Lowell, 1 University Ave., Lowell, MA 01854, USA

Advances in Biochemical Engineering/
Biotechnology, Vol. 41
Managing Editor: A. Fiechter

1 Microbial Formation of Poly(β-Hydroxyalkanoates)

1.1 Structure of PHA

A wide variety of different types of microorganisms are known to produce intracellular energy and carbon storage products which have been generally described as being poly(β-hydroxybutyrate), PHB [1–4]. The formation of this type of polymers is limited to prokaryotic organisms, whereas eukaryotic cells are not known to produce PHB. This particular polymer belongs to the family of poly(β-hydroxyalkanoates), (PHA), (**1**), which are usually formed as intracellular inclusions under stressed growth conditions; that is, in the presence of an excess of a carbon or energy source on the one hand and a limiting nutrient or growth factor on the other [1, 2, 5–8]. Because of these unbalanced growth conditions, reduction equivalents, which originate from metabolic oxidation processes, are stored in a water-insoluble, chemically and osmotically inert form of the following structure:

$$\left[-O-\overset{\overset{\displaystyle R}{|}}{\underset{*}{C}}H-CH_2-\overset{\overset{\displaystyle O}{\|}}{C}- \right]_m \qquad (1)$$

R = *n*-alkyl pendant group of variable chain length

HB, β-hydroxybutyrate where R = methyl
HV, β-hydroxyvalerate where R = ethyl
HC, β-hydroxycaproate where R = *n*-propyl
HH, β-hydroxyheptanoate where R = *n*-butyl
HO, β-hydroxyoctanoate where R = *n*-pentyl
HN, β-hydroxynonanoate where R = *n*-hexyl
HD, β-hydroxydecanoate where R = *n*-heptyl
HUD, β-hydroxyundecanoate where R = *n*-octyl
HDD, β-hydroxydodecanoate where R = *n*-nonyl

1.2 Occurrence of PHA in Microorganisms

The spectrum of PHA-producing microorganisms includes a variety of taxonomically different groups (Table 1). Most of the organisms are capable of accumulating PHA from 30 to 80% of their cellular dry weight. However, under specific conditions, *Alcaligenes eutrophus* N9A is known to contain 96% PHA [9].

These storage polyesters are also found in cyanobacteria such as *Aphanothece* or *Microcoleus* [10], but usually their PHA content is relatively low. Interstingly, enterobacteria are among the organisms which do not form PHA. However, it has been demonstrated recently that the PHB biosynthetic pathway from *A. eutrophus* can be cloned and expressed in *Escherichia coli* [11, 12]. Polymer contents of up 90% of the cellular dry weight were obtained from *E. coli*. In addition, it has been shown that PHA is a structural component of cell membranes of *E. coli* as well as *Haemophilus influenzae*, a strictly parasitic organism [13–16]. Besides some phototrophic microorganisms, *Clostridium* [17] and *Syntrophomonas* [18] are the only strictly anaerobic

Table 1. The accumulation of poly(β-hydroxyalkanoates) in a variety of microorganisms known to form intracellular storage products

Genus	Classification after Bergey's Manual[a]	Maximum PHA content (% dry wt)	PHA producing substrate	Reference
Acinetobacter	10	<1	Glucose	[105]
Alcaligenes	7	96	Fructose	[9]
Aphanothece	Cyanobacteria	<1	NS	[10]
Aquaspirillum	6	ND	NS	[106]
Azospirillum	6	75	Malate	[23]
Azotobacter	7	73	Glucose	[107]
Bacillus	15	25	Glucose	[28]
Beggiatoa	2	57	Acetate	[108]
Beijerinckia	7	38	Glucose	[109]
Caulobacter	4	36	Glucose/glutamate	[110]
Chloroflexus	1	<1	Yeast extract/glycylglycine	
Chlorogloea	Cyanobacteria	10	Acetate, CO_2	[111]
Chromatium	1	20	Acetate	[112]
Chromobacterium	8	37	Glucose/peptone	[113]
Clostridium	15	13	Tryptone/peptone/glucose	[114]
Derxia	7	26	Glucose	[109]
Ectothiorhodospira	1	ND	NS	[106]
Escherichia[b]	8	ND	Tryptone/yeast extract/glucose	[13]
Gamphosphaeria	Cyanobacteria	ND	ND	[115]
Haemophilus[b]	8	ND	Brain-heart-infusion	[13]
Halobacterium	13	38	Glucose	[20]
Hyphomicrobium	4	ND	Methanol, glucose	[22]
Lamprocystis	1	ND	NS	[106]
Lampropedia	10	ND	NS	[63]
Leptothrix	3	67	Pyruvate	[116]
Methylobacterium	7	47	Methanol	[117]
Methylocystis	ND	70	Methane	[118]
Methylosinus	7	25	Methane	[119]
Micrococcus	14	28	Peptone/tryptone	[120]
Microcoleus	Cyanobacteria	<1	NS	[10]
Microcystis	Cyanobacteria	ND	ND	[121]
Moraxella	10	ND	NS	[106]
Mycoplana	17	ND	Methanol	[122]
Nitrobacter	12	ND	NS	[106]
Nitrococcus	12	ND	NS	[106]
Nocardia	17	14	Butane	[65]
Oceanospirillum	6	ND	NS	[106]
Paracoccus	10	ND	NS	[106]
Photobacterium	8	ND	NS	[106]
Pseudomonas	7	67	Methanol	[27]
Rhizobium	7	57	Mannitol	[113]
Rhodobacter	1	80	Acetate	
Rhodospirillum	1	47	Acetate	[58]
Sphaerotilus	3	45	Glucose/peptone	[123]
Spirillum	6	40	Lactate	[21]
Spirulina	Cyanobacteria	6	CO_2	[124]
Streptomyces	17	4	Glucose	[125]
Syntrophomonas	9	30	Crotonate	[18]
Thiobacillus	12	ND	Glucose	[126]
Thiocapsa	1	ND	NS	[106]

Table 1. continued

Genus	Classification after Bergey's Manual[a]	Maximum PHA content (% dry wt)	PHA producing substrate	Reference
Thiocystis	1	ND	NS	[106]
Thiodictyon	1	ND	NS	[106]
Thiopedia	1	ND	NS	[106]
Thiosphaera	1	ND	Acetone, CO_2	[127]
Vibrio	8	ND	NS	[106]
Xanthobacter	7	ND	NS	[106]
Zoogloea	7	ND	Yeast extract/casamino acids	[128]

[a] after Ref. [106];
Group 1: Phototrophic bacteria;
Group 2: Gliding bacteria;
Group 3: Sheathed bacteria;
Group 4: Budding and/or appendaged bacteria;
Group 6: Spiral and curved bacteria;
Group 7: Gram-negative aerobic rods and cocci;
Group 8: Gram-negative facultative anaerobic rods;
Group 9: Gram-negative anaerobic bacteria;
Group 10: Gram-negative cocci and coccibacilli;
Group 12: Gram-negative chemolithotrophic;
Group 13: Archaebacteria;
Group 14: Gram-positive cocci;
Group 15: Endospore-forming rods and cocci;
Group 17: Actinomycetes.
[b] PHB found in cell membranes, not as intracellular inclusions;
ND Maximum PHA content not determined;
NS Substrate not specified

genera known to accumulate PHA in significant amounts. In addition, the presence of PHA has also been observed in a series of sulfate reducing bacteria [19]. From a phylogenetic point of view it might be noteworthy that PHA has been found both in Eubacteria and Archaebacteria [20].

1.3 Environmental Conditions Affecting PHA Formation

Generally, environmental conditions and physiological abilities of the bacteria control the quantitative formation of the storage polymer. Low concentrations or total absence of a variety of different nutrients can induce or stimulate the formation of PHA (Table 2). Usually, nitrogen-limiting conditions are chosen for the experimental work, because they are easily achieved by omitting ammonia from the growth medium. In addition, it has been found that PHA is also accumulated in certain organisms under oligotrophic growth conditions indicating a significant role of the polymer as part of the survival mechanism in nutrient-poor environments [21–23]. In this manner, the occurrence of PHA can serve as a marker of the nutritional status of a bacterial population [24–26], although limiting concentrations of calcium, cobalt, copper, molybdenum, sodium, and zinc seemed to have no influence on PHA formation by an organism such as *Pseudomonas* [27].

Table 2. List of limiting compounds leading to PHA formation

Compound	Organism	Reference
Ammonia	*Alcaligenes eutrophus*	[67]
	Alcaligenes latus	[129]
	Azospirillum brasiliense Cd	[23]
	Pseudomonas oleovorans	[5, 7]
	Pseudomonas cepacia	[130]
	Rhodospirillum rubrum	[58]
	Rhodobacter sphaeroides	
	Pseudomonas sp. K	[27]
	Methylocystis parvus	[118]
	Thiosphaera pantotropha	[127]
	Rhizobium ORS571	[131]
Carbon	*Spirillum* sp.	[31]
	Hyphomicrobium sp.	[22]
	Azospirillum brasiliense Cd	[23]
Iron	*Pseudomonas* sp. K	[27]
Magnesium	*Pseudomonas* sp. K	[27]
	Pseudomonas oleovorans	[132]
	Rhizobium ORS571	[131]
Manganese	*Pseudomonas* sp. K	[27]
Oxygen	*Azospirillum brasiliense* Cd	[23]
	Azotobacter vinelandii	[107]
	Azotobacter beijerinckii	[2]
	Rhizobium ORS571	[131]
Phosphate	*Rhodospirillum rubrum*	[58]
	Rhodobacter sphaeroides	
	Caulobacter crescentus	[110]
	Pseudomonas oleovorans	[132]
Potassium	*Bacillus thuringiensis*	[133]
Sulfate	*Pseudomonas* sp. K	[33]
	Pseudomonas oleovorans	[132]
	Rhodospirillum rubrum	[58]
	Rhodobacter sphaeroides	

Besides the importance as a source of energy and carbon under conditions of starvation, the intracellular presence of PHA (or PHB in particular) seems to play a significant role in the survival of the microorganism under other conditions of environmental stress such as when subjected to osmotic pressure, to desiccation, or to UV irradiation [23]. In general, cells containing this polymer have a higher survival rate than PHA-free cells. PHA also shows a physiological function during the sporulation in *Bacillus* [28] and the encystment in *Azotobacter* [2]. It has recently been suggested that PHB has a function in changing the barrier characteristics of cell membranes for DNA transfer in genetically competent bacteria. That is, besides its physiological role as a reserve or storage material, PHB has been found, therefore, to be a structural compound of the cell membrane [16].

1.4 PHA as Intracellular Inclusion Bodies

PHA is formed within the cell's cytoplasm as granular inclusions, which can be observed under the light microscope as refractile bodies [4, 29, 30], and which may or may not be crystalline in the living cell. Electron microscopic observations have shown that the granules are surrounded by a membrane which does not show the typical bilayered structure of a biomembrane [31, 32], and it was suggested that this membrane contains the PHA-synthetase or polymerase system [33–36]. Recently, the presence of a soluble PHB-synthetase in *A. eutrophus* was demonstrated which became granule-associated after the transition of the culture to nitrogen-limiting growth conditions [37].

Depending on the microorganisms, the PHA-depolymerase, which is responsible for the intracellular degradation of the polymer, may also be a part of this surrounding protein envelope [34]. However, some organisms show the presence of a soluble, cytoplasmatic depolymerase [38]. In any case, up to ten protein bands have been found in membranes of purified PHA granules from *P. oleovorans* when analyzed by polyacrylamide gel electrophoresis, presumably indicating the complex nature of the polymerase system which is located at the transition zone between the aqueous cytoplasm and the hydrophobic PHA granule.

2 Characterization and Properties of PHA

2.1 Physical and Chemical Characterization

PHB, which is the best-known member of the PHA series of polyesters, was discovered and initially described in 1925 by Lemoigne [39]. During the late 1950s and early 1960s metabolic pathways in the synthesis and degradation of PHB in *Bacillus* and *Rhodospirillum* was investigated [35, 40]. The first proposal for the use of PHB as a commercial plastic appeared when Baptist filed patents in which the application of PHB as an absorbable suture was suggested [41, 42]. Nevertheless, it was not until 1974 that Wallen and Rohwedder demonstrated the presence of other β-hydroxyalkanoate units than the β-hydroxybutyrate in the PHA found in sewage sludge [43]. Findlay and White [44] proposed, in 1983, to replace the term PHB by the more general description PHA when it was realized that the presence of pure homopolymers of β-hydroxybutyrate as intracellular storage products was the exception rather than the rule.

All of the PHA's found to date are basically linear polyesters of β-hydroxyalkanoic acids. The chirality of the β-carbon is responsible for the optical activity of these polymers, and all of the units are believed to exist in only the [*R*]-configuration. That is, in polymer chemistry terms, all of these polymers are perfectly isotactic. High molecular weights of up to 2×10^6 are found in solvent extracted polymer from different bacteria. This value corresponds to a degree of polymerization of approximately 20,000. Each PHA granule contains several thousand polymer chains [45].

It has been found that in the crystal lattice the conformational structure of PHB and copolymers of β-hydroxybutyrate and β-hydroxyvalerate is that of a right-handed helix with a two-fold screw axis and a fiber repeat of either 0.556 or 0.596 nm [46, 47]. The helix conformation is stabilized by carbonyl/methyl interactions [46] and represents one of the few exceptions of a helix found in nature which does not depend upon hydrogen bondings for its formation and stability. However, it has been demonstrated recently by X-ray diffraction studies of oriented films, that PHA with long

Table 3. Compositions of various PHA found in specific microbes and in environmental samples

	Type of β-hydroxyalkanoate units (number of carbon atoms in unit)[a]									Reference
	C-4	C-5	C-6	C-7	C-8	C-9	C-10	C-11	C-12	
R. rubrum	×	×	×	×						[58]
Rb. sphaeroides	×	×								
P. oleovorans	×	×	×	×	×	×	×	×	×	[6, 7]
P. cepacia	×	×								[130]
A. eutrophus	×	×								[72]
B. megaterium	×	×		×	×					[44]
Aphanothece	×	×								[10]
Microcoleus	×	×								[10]
Marine sediments	×	×	×	×	×					[44]
Sewage	×		×		×					[50]
Sewage sludge	×	×	×	×						[43]
Sewage sludge	×	×								[49]

[a] C-n, β-hydroxyalkanoate monomer unit with a chain length of *n* carbon atoms

pendant groups shows the typical d-spacings of comb-like polymers such as polymethacrylates, poly(vinyl ethers), or poly(vinyl esters) [48].

PHA extracted from sewage sludge [43, 49, 50], estuarine [51] and marine sediments [44], and compost heaps shows the presence of a variety of β-hydroxyalkanoate units indicating the presence of diverse microbial populations in these environments that are capable of producing different types of PHA. However, to date only a very limited number of organisms are known to incorporate repeating units longer than β-hydroxybutyrate into their storage polyesters (Table 3).

The physical and mechanical properties of these copolymers such as stiffness, brittleness, melting point, glass transition temperature, or resistance to organic solvents can change considerably as a function of the monomer composition [43, 48, 52]. Increasing the amount of β-hydroxyvalerate in β-hydroxybutyrate-β-hydroxyvalerate copolymers reduces the melting point from 180 °C for the PHB homopolymer to approximately 75 °C for a copolymer containing about 30 to 40 mole% β-hydroxyvalerate [43, 52]. PHA isolated from *P. oleovorans*, which contains long-chain repeating units, is soluble in acetone or ether, whereas PHB homopolymer is not [48].

2.2 Analytical Methods

Initially, the amount of PHA formed was determined by the digestion of cells containing granules with hypochlorite and then subsequently measuring the turbidity caused by the remaining polymer granules [28]. PHA amounts were also determined by gravimetric methods after extraction of the polymer with a solvent such as chloroform or propylene carbonate and subsequent precipitation of the polymer in ether [53, 54]. Another method for the quantitative determination of PHA was by UV spectrophotometric techniques after the acidic conversion of PHB to crotonic acid which was measured at 235 nm. However, this method is only useful for non-phototrophic organisms because the presence of pigments can interfere with the absorption of crotonic acid.

All of these methods do not allow for the determination of the monomer composition of the polymer, so for this purpose a quantitative determination of the types of monomers present other than β-hydroxybutyrate can be used, such as gas chromatography (GC) [55], high pressure liquid chromatography (HPLC) [56], and nuclear magnetic resonance spectroscopy (NMR) [57]. For the chromatographic techniques, the polymer is depolymerized to its equivalent monomer units by mild acidic or alkaline treatment, and the different repeating units are separated and identified. In contrast, the NMR method is, of course, non-destructive, and the polymer can be easily recovered.

Several methods have been applied to remove PHA quantitatively from bacterial cells, including solvent extraction [5, 48, 58, 59], hypochlorite treatment to destroy the cells [40, 60] followed by differential centrifugation [59, 61] and liquid/liquid extraction in an immiscible two-phase system [62]. It has been observed in several studies that the technique applied for cell disruption and for a subsequent extraction of the PHA were crucial for the retention of certain physical and chemical properties of the polyester [60]. The deleterious effect of hypochlorite on the molecular weight of the polymer has been demonstrated [60, 63]. The highest PHA yields and molecular weights of PHA isolated from *R. rubrum* were obtained by extraction of the polymer with hot chloroform as shown in Table 4. However, the relatively low polydispersity of two of the products in Table 4 indicates that most likely a loss of a lower molecular weight fraction occured during the extraction procedure. Mechanical destruction of whole cells using a French press also reduced the molecular weight significantly.

Table 4. Evaluation of different extraction and cell rupture methods for the PHA isolation from *Rhodospirillum rubrum*

	Chloroform extraction	Lysozyme, sonication	French press	Hypochlorite
PHA content (% dry wt)	13.6	13.6	13.6	13.6
Isolated PHA (% dry wt)	12.5	9.7	7.7	12.1
Recovery (%)	92	71	57	89
$M_w (\times 10^{-6})$ [a]	1.5	0.92	0.65	0.94
$M_n (\times 10^{-6})$ [b]	1.1	0.36	0.46	0.37
M_w/M_n [c]	1.4	2.6	1.4	2.5

[a] Weight-average molecular weight;
[b] Number-average molecular weight;
[c] Polydispersity

2.3 Functional PHA

There is only very little evidence for the presence of any type of functional groups in PHA produced in nature by bacterial synthesis. For instance, it has been demonstrated that *P. oleovorans* is capable of forming polyesters with unsaturated or alkene pendant groups when α-olefins are used as growth substrates [7, 64]. In that case, the double bond is found to exist as a terminal group in the pendant side chain. Hence, the double bonds of this type of PHA could be used as a reactive group for the chemical modification of the polymers. In addition, *Nocardia* sp. is able to form a copolymer of β-hydroxybutyrate and β-hydroxycrotonate as an intracellular storage material when the cells are grown on *n*-butane [65] leading to the incorporation of a double bond into the backbone carbon chain of the polymer. It has also been shown that chlorine is incorporated into the storage polymer of *A. eutrophus* when either 3-chloropropionic or 5-chlorovaleric acid are used as a carbon source [66–68]. The chloro group could also serve as a target for the chemical modification of PHA.

Furthermore, quite recently, unusual copolymers containing 3-hydroxybutyrate and 4-hydroxybutyrate units were obtained, as well as copolymers of 3-hydroxybutyrate, 3-hydroxyvalerate, and 5-hydroxyvalerate from *A. eutrophus* indicating an unusual flexibility of this organism in its ability to produce polymers which do not contain exclusively poly(β-hydroxyalkanoate) units [57, 69].

3 PHA as Biodegradable Plastics

3.1 Industrial Production of PHA

It has become of considerable industrial interest and of environmental importance to evaluate PHA as polyesters for the use in either biodegradable and biocompatible plastics for a wide range of possible applications as indicated in Table 5. Early investigations of PHA granules by electron microscopy after freeze-etching showed that the polymer in the granule underwent a cold drawing process indicating the plastic nature of the polyester [31] and suggesting that it could be processed as a conventional thermoplastic. In addition to its potential as plastic material, PHA represent also a useful source for stereoregular compounds which can serve as chiral precursors for the chemical synthesis of optically active substances, particularly in the synthesis of certain drugs or insect pheromones [70–72]. These substances are biologically active only in the correct stereochemical configuration.

An example of a useful PHA polyester is the one which is now being produced industrially, on a fairly large scale, by a subsidiary of ICI Ltd. in Great Britain using

Table 5. Practical applications of PHA [68, 75–77, 104]

Medical applications:

- Surgical pins, sutures, staples, and swabs
- Wound dressing
- Blood vessel replacements
- Bone replacements and plates
- Stimulation of bone growth by piezoelectric properties
- Biodegradable carrier for long term dosage of drugs and medicines

Industrial applications:

- Biodegradable carrier for long term dosage of herbicides, fungicides, insecticides, or fertilizers
- Packaging containers, bottles, wrappings, bags and films
- Disposable items such as diapers or feminine hygiene products

A. eutrophus H16 [68, 73]. Either PHB homopolymer or copolymers of β-hydroxybutyrate and β-hydroxyvalerate can be formed by these cells depending on the substrate or substrate mixtures used for growth and polymer production [74]. These polymers are commercially available under the trade name "Biopol".

For industrial applications it is desirable to control the incorporation of different repeating units into the polymer in order to produce polyesters with specific material characteristics because their physical and chemical characteristics depend strongly on copolymer composition. "Tailor-made" copolymers can be made for this purpose by the use of controlled growth conditions. If a defined mixture of nutrients for a certain type of microorganisms is supplied for growth, a defined and reproducible copolymer is formed.

Biochemically, there are two different ways of achieving polymer formation in microbes. As indicated in Fig. 1 the growth substrate or carbon source can be metabolized to form both microbial biomass and storage pôlymer simultaneously, as demonstrated by the formation of various PHA copolyesters in *R. rubrum* and *P. oleovorans* [6, 48, 58]. During cell growth, however, part of the polymer-forming potential is lost because of the utilization of the substrate to maintain the cell's metabolism. The second possible method of industrial polymer production is a serial process where microorganisms are first grown on a carbon source to obtain a large amount of biomass, then the medium is depleted of an essential nutrient and a polymer-forming substrate is added. The latter is converted directly to polymers and essentially only little additional cell growth occurs. This approach is used for the large scale PHA production by *A. eutrophus* [57, 66, 67, 69].

PHB homopolymer shows similiarities in its physical properties and even in its molecular structure to the isotactic polypropylene (PP). For the latter, both polymers have a methyl pendant group attached to the main chain in a single configuration. Table 6 shows a comparison of several of the material properties [46, 75, 76]. The main difference between the two, however, is the biodegradability of PHB compared to the insignificant degradation of PP. A further difference, which might be important in its practical application, is the density of the two plastic materials. Because of its

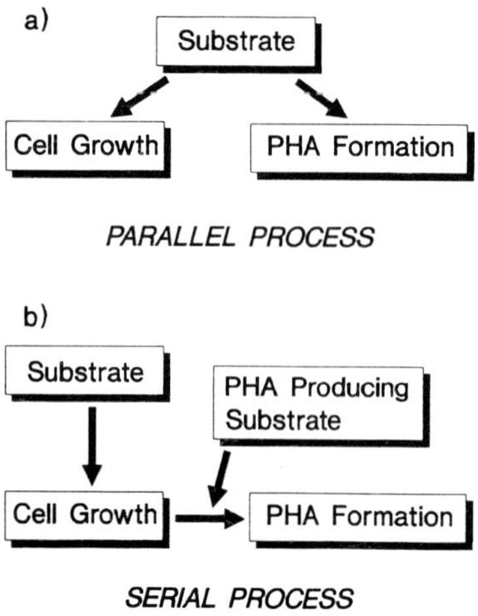

Fig. 1 a, b. Microbial PHA formation: **a)** a substrate is utilized simultaneously for both cell growth and PHA formation; **b)** a polymer producing substrate is added to the culture after growth

Table 6. Chemical and physical properties of polypropylene (PP) and poly(β-hydroxybutyrate) (PHB) [46, 74–76]

Parameter	PP	PHB
Melting point T_m [°C]	171–186	171–182
Glass transition temperature T_g [°C]	−15	5–10
Crystallinity [%]	65–70	65–80
Density [g cm^{-3}]	0.905–0.94	1.23–1.25
Molecular weight M_w($\times 10^{-5}$)	2.2–7	1–8
Molecular weight distribution	5–12	2.2–3
Flexural modulus [GPa]	1.7	3.5–4
Tensile strength [MPa]	39	40
Extension to break [%]	400	6–8
UV resistance	poor	good
Solvent resistance	good	poor
Oxygen permeability [cm^3 m^{-2} atm^{-1} d^{-1}]	1700	45
Biodegradability	+	—
Approx. U.S. annual production [Mio. t]	1.8	ND[a]

[a] not determined

high density, PHB does not float in a aquatic ecosystem, but PP does. Therefore, once discarded, plastic articles made from PHB will sink and will be degraded in the surface sediments by biogeochemical mechanisms.

3.2 Biodegradation of PHA

In general, a series of parameters can influence the rate of biodegradation and the life span of a plastic material in nature, including the type of environment, the presence of a microbial population and its microbial activity, the availability of water, the temperature, the section thickness of the plastic material, its surface texture, its porosity, and the presence of second components in the plastic, such as fillers or coloring agents. As examples, the effect of different environments and environmental conditions on the degradation rate of PHB homopolymer films has been determined [74, 77]. In humid air, degradation was negligible, whereas in anoxic sewage 100% degradation of a 1 mm thick film occurred over six weeks. In oxic sewage or soil, 60 or 75 weeks, respectively, are required for a complete degradation. In seawater at 15 °C, the film was degraded within 350 weeks.

Two mechanisms are responsible for the degradation of PHA. Under sterile or aseptic conditions, PHA is degraded by a hydrolylytic mechanism, especially at high pH [78, 79]. This type of degradation is important for medical applications, such as the use of PHA in drug release carriers or surgical sutures. In natural environments, the polymer is degraded enzymatically by the action of a depolymerase or esterase.

Several microorganisms such as *Alcaligenes faecalis* [80] *Pseudomonas lemoignei* [81–83], *Penicillium simplicissimum* [84], or *Eupenicillium* sp. [84] are known to degrade PHB by the secretion of an esterase to catalyze the hydrolysis of the polymer in order to generate organic compounds which can be used as a carbon source for growth. PHB can aso be degraded by mixed bacterial cultures which are obtained from soil

[85]. In contrast to PHB, PHA's with longer chain length pendant groups should undergo slower biodegradation rates in environments such as sewage sludge or sediment for the following reasons: (a) Low molecular weight carbon compounds are mineralized more rapidly, whereas more complex structures have much lower rates of mineralization [86, 87], and (b) long-chain repeating units increase the hydrophobicity of the polymer, and a high hydrophobicity generally inhibits or prevents microbial growth on the polymer surface. Therefore, it is likely that PHA with long pendant groups are degraded more slowly than PHB so that the lifetime of a plastic article made from PHA derived from *P. oleovorans* may be prolonged.

Table 7 shows the biodegradability of a group of PHA's which differ in composition, stereochemistry and molecular weight. In general the initial molecular weight has no significant influence on the biodegradability of the polymer, and a high molecular weight PHB homopolymer is degraded as rapidly as a low molecular weight polymer. On the other hand, the degradation is prevented when units with the [*S*]-configuration are present, indicating the stereoselectivity of the esterase enzyme. Copolymers of β-hydroxybutyrate and β-hydroxyvalerate are readily degraded independent of the amount of β-hydroxyvalerate incorporated in the polymer. PHA polyester with β-hydroxyheptanoate as the major repeating unit was degraded as well. In contrast, no degradation of the polymer was detected when β-hydroxynonanoate was the main repeating unit. These results suggest that by specific control of the unit present, the biodegradability and the degradation rates of the PHA polymers could be controlled.

The very wide range of microbes known to produce PHA allows the selection of organisms which are capable of forming specific polyesters under defined conditions. On an industrial scale, the use of photosynthetic bacteria, for instance, would harness sunlight as an energy source for the production of these polyesters. Because of their metabolic flexibility, organisms like *Pseudomonas* have the potential ability to produce a whole new variety of useful plastics all of which would be biodegradable.

The early investigations described in this review indicate that there is a wide versatility in nature of many different types of polyesters being formed, many of which could

Table 7. Aerobic degradation of various types of PHA films by a mixed bacterial culture enriched from freshwater sediments

Type[a]	Configuration	Composition of PHA (%)	M_w	Degradation[b]
PHB	[*R*]	100	150'000	+
PHB	[*R*]	100	800'000	+
PHB	[*R*, *S*] stereoblock	100	ND[c]	—
PHB	[*R*, *S*] atactic	100	ND	—
PHB	[*S*]	100	ND	—
PHB-*co*-HV	[*R*]	83/17	150'000	+
PHB-*co*-HV	[*R*]	80/20	ND	+
PHB-*co*-HV	[*R*]	70/30	ND	+
PHH-*co*-HV	[*R*]	94/4	360'000	+
PHN-*co*-HH	[*R*]	59/32	190'000	—

[a] structure according to formula (1);
[b] +: polymer degraded; —: polymer not degraded;
[c] not determined

be used as substitutes for petrochemically based plastics. To achieve this goal, however, major efforts are needed to scale up laboratory processes to an industrial production level and to expand the number of available PHA forming microorganisms in order to obtain a wide range of microbially produced PHA.

4 Biodegradable Plastics and Solid Waste Disposal

4.1 Ecological Aspects

The increased sensitivity to ecological problems will have a major impact in the future on the disposal of plastic articles. In time, the plastics industry may be forced to explore the production of biodegradable polymers as an alternative to traditional plastics. Indeed, because of the increasing legislative efforts to ban the disposal of non-degradable plastic articles, there is already considerable pressure on the plastics industry to substitute existing non-degradable plastics for biodegradable materials. That is, one of the greatest virtues of plastics, their durability, is becoming its greatest problem [88]!

The tremendous production and use of plastic materials around the world create problems of massive waste disposal, both planned and random. For example, several hundred thousand tons of plastic are discarded into marine environments each year [89–91], and this waste material accumulates in certain regions of the ocean [94]. As a result, it has been estimated that one million marine animals are killed every year either by choking on floating plastic items which were mistaken for food sources or by entanglement in non-degradable plastic debris [89–92]. One effect of the former is that the ingestion of plastic particles reduces the eating capacity of a sea bird food thus reducing the formation of fat deposits and leading to reduced fitness [93].

4.2 Disposal of Plastics Waste

At present, the only general methods of coping with the disposal of non-degradable plastics are in the use of landfills whose capacity is rapidly being exhausted [94, 95].

Incineration of plastics is potentially dangerous and can be expensive. During the combustion of plastic waste, hydrogen cyanide can be formed from acrylonitrite based plastics and nylon, and hydrogen chloride is released from PVC [96]. These air pollutants have to be removed by scrubbing devices.

Post-user recycling of plastic materials is of limited value and use. The sorting of the wide variety of discarded plastic items is a major problem for recycling these materials [97]. Also, the presence of additives, such as pigments, coatings, fillers, and other reinforcements, limits the post-user recovery significantly as well as the subsequent use of these materials. The recycling process changes the material properties in an adverse manner, thus limiting their application range. In addition, the virgin polymers, in many cases, are still less expensive than recycled materials [97].

A partial solution to the increase in plastic litter and the waste disposal problem is the development and the industrial production of degradable materials [90]. Several technologies have been used to build the qualities of biodegradability and photodegradability into certain materials. Besides the utilization of photodegradable plastics, biodegradable polymers would certainly contribute to a reduction of the currently rising plastic waste heaps. Plant derived starch has been used to produce biodegradable plastic articles, such as pharmaceutical capsules, by blow molding processes [90]. Starch has also been used as a natural filler or an additive in a basically non-degradable matrix such as polyethylene. When utilized by microorganisms the starch will leave a porous structure that disintegrates due to its mechanical instability [90]. Further sources of biodegradable materials are poly(lactic acid), poly(malic acid), or poly(ε-caprolactones) [98–102], which are synthesized chemically. In contrast, PHA's are produced microbially from renewable, plant derived feedstocks [68, 76, 103, 104]. In addition, because it can be processes by traditional techniques used in the plastics' industry such as injection molding, extrusion, blow molding, melt casting, or spinning, PHA has the potential to become an important source material for commodity plastics which are biodegradable [103, 104].

However, the biodegradable plastics currently available can only be used for certain specific applications. High performance applications, such as in high temperature or impact resistant plastics, cannot be replaced at this time, and there are other problems related to the use of biodegradable materials. A prolonged contact with biodegradable plastics used as packaging materials could conceivably cause the contamination of food by undesirable microbes which could grow on the plastic. The uncontrolled diposal of some biodegradable plastic items could also cause the eutrophication of certain aquatic ecosystems, because the material would serve as easily mineralizable organic matter.

In any case, the acceptance of such materials is expected to be very gradual, and it has been predicted that in the year 2002 only 3% of the estimated annual 15 million t of plastic-packaging waste will be biodegradable. Instead, the major portion, 42.5%, is expected to be recycled, 37% buried in landfills, and 17.5% incinerated [97].

5 Acknowledgement

This work was supported by the Office of Naval Research, Molecular Biology Program under Grant No. N00014-86K-0369.

6 Abbreviations

PHA:	Poly(β-hydroxyalkanoates)
PHB:	Poly(β-hydroxybutyrate)
PHB-*co*-HV:	Poly(β-hydroxybutyrate-*co*-hydroxyvalerate)
PHH-*co*-HV:	Poly(β-hydroxyheptanoate-*co*-hydroxyvalerate)
PHN-*co*-HH:	Poly(β-hydroxynonanoate-*co*-hydroxyheptanoate)

7 References

1. Byrom D, (1987) Trends in Biotechnol. 5: 246
2. Dawes EA, (1974) The role and regulation of poly-β-hydroxybutyrate as a reserve in microorganisms. In: Mano EB (ed) Proceedings of the International Symposium on Macromolecules, Rio de Janeiro, July 26–31. Elsevier, Amsterdam, p 433
3. Schlegel HG, Gottschalk G (1962) Angew. Chem. 74: 342
4. Shively JM (1974) Ann. Rev. Microbiol. 28: 167
5. Brandl H, Gross RA, Lenz RW, Fuller RC (1988) Appl. Environ. Microbiol. 54: 1977
6. Brandl H, Gross RA, Lenz RW, Fuller RC (1988) Abstracts of the annual meeting of the American Society for Microbiology, Miami Beach, p 272
7. Lageveen RG, Huisman GW, Preusting H, Ketelaar P, Eggink G, Witholt B (1988) Appl. Environ. Microbiol. 54: 2924
8. Suzuki T, Yamane T, Shimizu S (1986) Appl. Microbiol. Biotechnol. 24: 370
9. Pedrós-Alió C, Mas J, Guerrero R (1985) Arch. Microbiol. 143: 178
10. Capon RJ, Dunlop RW, Ghisalberti EL, Jefferies PJ (1983) Phytochemistry 22: 1181
11. Slater CS, Voige WH, Dennis DE (1988) J. Bacteriol. 170: 4431
12. Schubert P, Steinbüchel A, Schlegel HG (1988) J. Bacteriol. 170: 5837
13. Reusch RN, Hiske T, Sadoff HL (1986) J. Bacteriol. 168: 553
14. Reusch RN, Sadoff HL (1983) J. Bacteriol. 156: 778
15. Reusch RN, Hiske T, Sadoff HL, Harris R (1987) Can. J. Microbiol. 33: 435
16. Reusch RN, Sadoff HL (1988) Proc. Natl. Acad. Sci. USA 85: 4176
17. Emeruwa AC, Hawirko RZ (1973) J. Bacteriol. 116: 989
18. Amos DA, McInerey MJ (1987) Abstracts of the annual meeting of the American Society for Microbiology, Atlanta, p 174
19. Pfennig N (personal communication)
20. Fernandez-Castillo R, Rodriguez-Valera F, Gonzales-Ramos J, Ruiz-Berraquero F (1986) Appl. Environ. Microbiol. 51: 214
21. Benedict CV, Cameron JA, Huang SJ (1983) J. Appl. Polym. Sci. 28: 335
22. Nitikin DI, Pitruyk IA, Zagreba ED, Ginovska MK, Yacobson YO, Fetisova MB (1986) Microbiology 55: 509
23. Tal S, Okon Y (1985) Can. J. Microbiol. 31: 608
24. Nickels JS, King JD, White DC (1979) Appl. Environ. Microbiol. 37: 459
25. Tunlid A, Baird BH, Trexler MB, Olsson S, Findlay RH, White DC (1985) Can. J. Microbiol. 31: 1113
26. White DC, Smith GA, Gehron MJ, Parker JH, Findlay RH, Martz RF, Fredrickson HL (1983) Dev. Industr. Microbiol. 24: 201
27. Suzuki T, Yamane T, Shimizu S (1986) Appl. Microbiol. Biotechnol. 23: 322
28. Williamson DH, Wilkinson JF (1958) J. Gen. Microbiol. 19: 198
29. Dawes EA, Senior PJ (1973) Adv. Microb. Physiol. 10: 135
30. Merrick JM (1978) Metabolism of reserve materials. In: Clayton RK, Sistrom WR (eds) The Photosynthetic Bacteria, Plenum, New York, p 199
31. Dunlop WF, Robards AW (1973) J. Bacteriol. 114: 1271
32. Ellar D, Lundgren DG, Okamura K, Marchessault RH (1968) J. Molec. Biol. 35: 489
33. Merrick JM (1988) Polym. Prepr. 29 (1): 586
34. Merrick JM, Doudoroff M (1961) Nature 189: 890
35. Merrick JM, Lundgren DG, Pfister RM (1965) J. Bacteriol. 89: 234
36. Merrick JM, Yu CF (1966) Biochemistry 5: 3563
37. Haywood GW, Anderson AJ, Dawes EA (1989) FEMS Microbiol. Lett. 57: 1
38. Dawes EA (1986) Microbial energetics. Blackie, Glasgow, London
39. Lemoigne M (1925) Ann. Inst. Pasteur 39: 144
40. Macrae RM, Wilkinson JF (1958) J. Gen. Microbiol. 19: 210
41. Baptist JN (1959) US patent 3036959
42. Baptist JN (1960) US patent 3044942
43. Wallen LL, Rohwedder WK (1974) Environ. Sci. Technol. 8: 576
44. Findlay RH, White DC (1983) Appl. Environ. Microbiol. 45: 71

45. Cornibert J, Marchessault RH, Benoit H, Weill G (1970) Macromolecules 3: 741
46. Bloembergen S (1987) Characterization of bacterial poly(β-hydroxybutyrate-*co*-β-hydroxyvalerate) and synthesis of analogues via a non-biochemical approach. Thesis, University of Waterloo, Canada
47. Okamura K, Marchessault RH (1967) X-ray structure of poly-β-hydroxybutyrate. In: Ramachandran GN (ed) Conformation in Biopolymers, vol. 2. Academic, New York, p 709
48. Gross RA, DeMello C, Lenz RW, Brandl H, Fuller RC (1989) Macromolecules 22: 1106
49. Comeau Y, Hall KJ, Oldham WK (1988) Appl. Environ. Microbiol. 54: 2325
50. Odham G, Tunlid A, Westerdahl G, Marden P (1986) Appl. Environ. Microbiol. 52: 905
51. Herron JS, King JD, White DC (1987) Appl. Environ. Microbiol. 35: 251
52. Bluhm TL, Hamer GK, Marchessault RH, Fyfe CA, Veregin RP (1986) Macromolecules 19: 2871
53. Schlegel HG (1962) Arch. Microbiol. 42: 110
54. Lafferty RM, Heinzle E (1978) US Patent 4101533
55. Braunegg G, Sonnleitner B, Lafferty RM (1978) Eur. J. Appl. Microbiol. Biotechnol. 6: 29
56. Karr DB, Waters JK, Emerich DW (1983) Appl. Environ. Microbiol. 46: 1339
57. Doi Y, Kunioka M, Nakamura Y, Soga K (1986) Macromolecules 19: 2860
58. Brandl H, Knee EJ Jr, Fuller RC, Gross RA, Lenz RW (1989) Int. J. Biol. Macromol. 11: 49
59. Griebel R, Smith Z, Merrick JM (1968) Biochemistry 7: 3676
60. Griebel R, Merrick JM (1971) J. Bacteriol. 108: 782
61. De Smet MJ, Eggink G, Witholt B, Kingma J, Wynberg H (1983) J. Bacteriol. 154: 870
62. Hofsten VB, Baird GD (1962) Biotechnol. Bioengin. 6: 403
63. Lundgren DG, Alper R, Schnaitman C, Marchessault RH (1965) J. Bacteriol. 89: 245
64. Witholt B, Lageveen RG, Huisman GW, Preusting H, Nijenhuis A, Kingma J (1988) Polym. Prepr. 29 (1): 592
65. Davis JB (1964) Appl. Microbiol. 12: 301
66. Doi Y, Kunioka M, Nakamura Y, Soga K (1987) Macromolecules 20: 2988
67. Doi Y, Tamaki A, Kunioka M, Soga K (1987) Makromol. Chem., Rapid Comm. 8: 631
68. Holmes PA, Wright LF, Collins SH (1981) Eur. Pat. Appl. 0052459
69. Kunioka M, Tamaki A, Doi Y (1989) Macromolecules 22: 694
70. Seebach D, Züger MF (1982) Helv. Chim. Acta 65: 495
71. Seebach D, Züger MF (1985) Tetrahedron Lett. 25: 2747
72. Sonnet PE (1988) Chemtech 18: 94
73. Holmes PA (1985) Phys. Technol. 16: 32
74. Winton JM (1985) Chem. Week Aug. 28: 55
75. Howells ER (1982) Chem. Ind. 1982: 508
76. King PP (1982) J. Chem. Technol. Biotechnol. 32: 2
77. Uttley NL (1986) Appl. Biotechnol., Proc. BIOTECH '86, vol 1, p 171
78. Holland SJ, Jolly AM, Yasin M, Tighe BJ (1987) Biomaterials 8: 289
79. Miller ND, Williams DF (1987) Biomaterials 8: 129
80. Tanio T, Fukui T, Shirakura Y, Saito T, Tomita K, Kaiho T, Masamune S (1982) Eur. J. Biochem. 124: 71
81. Delafield FP, Doudoroff M, Palleroni NJ, Lustry CJ, Contopoulos R (1965) J. Bacteriol. 90: 1455
82. Lusty CJ, Doudoroff M (1966) Proc. Natl. Acad. Sci. USA 56: 960
83. Nakayama K, Saito T, Fukui T, Shirakura Y, Tomita K (1985) Biochim. Biophys. Acta 827: 63
84. McLellan DW, Halling PJ (1988) FEMS Microbiol. Lett. 52: 215
85. Chowdhury AA (1963) Arch. Microbiol. 47: 167
86. Cranwell PA (1976) Organic geochemistry in lake sediments. In: Nriagu JO (ed) Environmental Biochemistry, vol. 1. Ann Arbor Science, Michigan, p 75
87. Fenchel T, Blackburn TH (1979) Bacteria and Mineral Cycling. Academic, London, New York, San Francisco
88. Anonymous (1988) Mod. Plast. May 1988: 148
89. Bean MJ (1987) Mar. Poll. Bull. 18: 357
90. Leaversuch R (1987) Mod. Plast. Aug. 1987: 52
91. Pruter AT (1987) Mar. Poll. Bull. 18: 305
92. Wilber RJ (1987) Oceanus 30 (3): 61

93. Ryan PG (1988) Mar. Poll. Bull. 19: 125
94. O'Leary PR, Walsh PW, Ham RK (1988) Sci. Am. 259: 19
95. Crawford M (1988) Science 241: 411
96. Huffman GL, Keller DJ (1973) The plastics issue. In: Guillet J (ed) Polymers and Ecological Problems. Plenum, New York, London, p 155
97. Leaversuch R (1988) Mod. Plast. June 1988: 65
98. St. Pierre T, Chiellini E (1986) J. Bioact. Biocomp. Polym. 1: 467
99. St. Pierre T, Chiellini E (1987) Bioact. Biocomp. Polym. 2: 4
100. Benedict CV, Cameron JA, Huang SJ (1983) J. Appl. Polym. Sci. 28: 335
101. Benedict CV, Cook WJ, Jarrett P, Cameron JA, Huang SJ, Bell JP (1983) J. Appl. Polym. Sci. 28: 327
102. Cook WJ, Cameron JA, Bell JP, Huang SJ (1981) J. Polym. Sci: Polym. Lett. Ed. 19: 159
103. Mann S, Calvert PD (1987) Trends in Biotechnol. 5: 309
104. Uttley NL (1985) Manuf. Chem. Oct. 1985: 63
105. Martin DW, Seaton HJ, Stewart JR (1988) Abstracts of the annual meeting of the American Society for Microbiology, Miami Beach, p 216
106. Buchanan RE, Gibbons NE (eds) (1974) Bergey's Manual of Determinative Bacteriology. Williams and Wilkins, Baltimore
107. Ward AC, Rowley BI, Dawes EA (1977) J. Gen. Microbiol. 102: 61
108. Güde H, Strohl WR, Larkin JM (1981) Arch. Microbiol. 129: 357
109. Stockdale H, Ribbons DW, Dawes EA (1968) J. Bacteriol. 95: 1789
110. Poindexter JS, Eley LF (1983) J. Microbiol. Meth. 1: 1
111. Carr NG (1966) Biochim. Biophys. Acta 120: 308
112. Schlegel HG (1962) Arch. Microbiol. 42: 110
113. Forsyth WG, Hayward AC, Roberts JB (1958) Nature 182: 800
114. Emeruwa AC, Hawirko RZ (1973) J. Bacteriol. 116: 989
115. Cmiech HA, Leedale GF, Reynolds CS (1987) Br. Phycol. J. 22: 339
116. Adams LF, Ghiorse WC (1986) Arch. Microbiol. 145: 126
117. Powell KA, Collinson BA, Richardson KR (1983) Eur. Pat. Appl. 0015669
118. Asenjo JA, Suk JS (1986) J. Ferment. Technol. 64: 271
119. Dalton H, Stirling DI (1982) Phil. Trans. R. Soc. Lond. 297: 481
120. Sierra G, Gibbons NE (1962) Can. J. Microbiol. 8: 249
121. Reynolds CS, Jaworski GHM, Cmiech HA, Leedale GF (1981) Phil. Trans. R. Soc. Lond. 293: 419
122. Sonnleitner B, Heinzle E, Braunegg G, Lafferty RM (1979) Eur. J. Appl. Microbiol. Biotechnol. 7: 1
123. Stokes JL, Powers MT (1967) Arch. Microbiol. 59: 295
124. Campbell J, Stevens SE, Balkwill DL (1982) J. Bacteriol. 149: 361
125. Kannan LV, Rehacek Z (1970) Ind. J. Biochem. 7: 126
126. Wang WS, Lundgren DG (1969) J. Bacteriol. 97: 947
127. Bonnet-Smits EM, Robertson LA, Van Dijken JP, Senior E, Kuenen JG (1988) J. Gen. Microbiol. 134: 2281
128. Fukui T, Yoshimoto A, Matsumoto M, Hosokawa S, Saito T, Nishikawa H, Tomita K (1976) Arch. Microbiol. 110: 149
129. Lafferty RM, Braunegg G, Korneti L, Strempfl B, Bogensberger B, Korsatko W, Wabnegg B (1984) Poly-D-(—)-3-hydroxybutyric acid (poly-HB): Biotechnological production and polymer application. Third European Congress on Biotechnology, vol 1, Verlag Chemie, Weinheim, p 521
130. Ramsay BA, Ramsay JA, Cooper DG (1989) Appl. Environ. Microbiol. 55: 584
131. De Vries W, Stam H, Duys JG, Ligtenberg AJM, Simons LH, Stouthamer AH (1986) Antonie van Leeuwenhoek J. Microbiol. 52: 85
132. Lageveen R (1986) Oxidation of aliphatic compounds by *Pseudomonas oleovorans*: Biotechnological applications of the alkane-hydroxylase system. Thesis, University of Groningen, The Netherlands
133. Wakisaga Y, Masaki E, Nishimoto Y (1982) Appl. Environ. Microbiol. 43: 1473

Glycerol

Gopal Prasad Agarwal*
Biochemical Engineering Research Center, Indian Institute of Technology, New Delhi-110016, India

1 Introduction . . . 96
1.1 History . . . 96
1.2 Definition and Properties . . . 96
1.3 Future Prospects . . . 97
2 Process Description . . . 97
2.1 Chemical Routes of Glycerol Production . . . 98
2.2 Biochemical Routes of Glycerol Production . . . 99
2.2.1 Sulfite-/Alkali-Steered Yeast Process . . . 100
2.2.2 Bacterial Process . . . 103
2.2.3 Osmotolerant Yeast Process . . . 105
2.2.4 Algal Cultivation Process . . . 107
3 Downstream Processing . . . 110
3.1 Glycerol via Chemical Processes . . . 110
3.2 Glycerol via Yeasts and Bacterial Processes . . . 111
3.3 Glycerol via Algal Cultivation . . . 112
4 Biosynthesis of Glycerol . . . 113
4.1 Normal Yeast Metabolism . . . 114
4.2 Osmotolerant Yeast Metabolism . . . 115
4.3 Bacterial Metabolism . . . 117
4.4 Osmoregulatory Mechanism of Algae . . . 119
5 Comparative Study of the Processes . . . 120
5.1 Raw Material . . . 120
5.2 Energy Consumption . . . 121
5.3 Productivity, Yield and Final Concentration . . . 122
6 Summary . . . 124
7 Acknowledgements . . . 126
8 References . . . 126

Glycerol is traditionally produced as a by-product of soap and fatty acids industries. The demand for glycerol has always exceeded the supply from these industries so the excess demand has been met by chemical synthesis from propylene for the last several decades. Though glycerol production has a long history (dating back to World War I) of being produced via a biochemical route, yet it is not sufficiently developed to compete with the chemical route. In the present review a case has been made to produce glycerol via any of the several known biochemical routes: a) Sulfite-Alkali-Steered Yeast Process b) Bacterial Process c) Osmotolerant Yeast Process d) Algal Cultivation Process. The possible reasons for these processes not being able to compete with chemical processes are critically reviewed. The literature on downstream processing of any of the biochemical processes is quite limited and more investigations are required into this aspect to make these processes viable. The biosynthesis mechanism of glycerol production in the organisms is summarized and the need to look into some of the fundamental aspects of glycerol synthesis in an osmotolerant yeast emphasized. The comparison between the various processes is made wherever possible.

* A part of the chapter was rewritten when visiting M.I.T., Cambridge, USA

Advances in Biochemical Engineering/
Biotechnology, Vol. 41
Managing Editor: A. Fiechter

1 Introduction

1.1 History

Glycerol, a very important polyhydroxy alcohol, has been traditionally produced as a by-product of soap and fatty acids industry [1]. It was produced for the first time on a large scale using the sulfite-steered yeast process during World War I when demand for glycerol as explosives for war time use exceeded the supply from the soap industry [2]. However, war-time process technology could never adopt itself to the peace-time competition from chemically synthesized process developed after World War II. One of the earlier uses of glycerol had been in the mining and ordnance industries. Since then, glycerol has found wider applications in many chemical and plastic industries and its use as an explosive had dwindled to less than 2.5% by 1980 in USA [3]. The largest amount goes into the manufacture of synthetic resins and ester gums, drugs, cosmetics, and tooth pastes. Tobacco processing and foods also consume large amounts either as glycerol or glycerides. The use of glycerol as an explosive, which was one of the main uses during World War I, has dwindled in face of better explosives available today. Though the demand for glycerol has stabilized in the USA in the last decade, demand is likely to go up in developing countries as their standard of living improves. In the late 1940s, developed countries started producing glycerol via chemical synthesis routes; however, many developing countries continue to depend upon the glycerol supply from the soap industry and when demand for glycerol exceeds the supply, it is met by importing from developed countries. Production of soap in the future is not expected to increase in developing countries as the use of detergents becomes more widespread. Therefore, it becomes imperative to look for a new process for glycerol production especially in developing countries. Since molasses is cheaply available in those countries, production of glycerol via a biochemical route may provide the additional glycerol for the growing demand. The chemical route may not be a viable process in the long run as most of the developing countries are also importers of crude petroleum, a major raw material for the chemical route. Even for developed countries, the biochemical route may compete with the chemical route as the former is based on renewable raw materials while the latter is using up the limited depleting reserves of crude petroleum.

1.2 Definition and Properties

The term glycerol applies only to the pure chemical compound 1,2,3-propanetriol $CH_2OH \cdot CHOH \cdot CH_2OH$. The term glycerine applies to the purified commercial products normally containing > 95% of glycerol. Several grades of glycerine are available commercially which differ somewhat in their glycerol content and in other characteristics such as color, odor, and trace impurities [3]. The spelling 'glycerine' should be discouraged since the ending -ine is chemically indicative of organic bases such as amines, and does not apply to glycerine, therefore, the word glycerol is used throughout this chapter. Glycerol is a clear, water-white, viscous, hygroscopic liquid with a sweet taste at ordinary room temperatures above its melting point. Glycerol

2.2.3 Osmotolerant Yeast Process

A possible solution to the problems raised by the presence of steering agents in the complex medium of ordinary yeast process resulting in difficult glycerol recovery was suggested by Nickerson and Carroll [31]. They discovered that glycerol was produced by an osmotolerant yeast, at that time classified as *Zygosaccharomyces acidifaciens*, which did not require bisulfites, sulfites or any other steering agents. The yeast used by Nickerson and Carroll was eventually reclassified in the genus *Saccharomyces* and is currently classified as *S. bailii* [32]. The report of Nickerson and Carroll stimulated investigation of the biosynthetic properties of the osmophilic yeasts in the hope that an organism producing a good yield of glycerol without steering agents might be found. However, the yields of glycerol actually recovered were considerably lower than were indicated by the periodate-chromotropic acid analysis [33, 34]. Later it was discovered that osmophilic yeasts were capable of producing other polyhydroxy alcohols (in short polyols) like erythritol, D-arabitol (also known as arabinitol) and mannitol which were also detected as glycerol by periodate-chromatographic acid analysis.

In the 1950s, other strains of sugar-tolerant yeasts were studied at the Prairie Regional Laboratory of the National Research Council of Canada at Saskatoon in Saskatchewan [35, 36] and at the Forest Products Laboratory, University of Wisconsin at Madison [37, 38]. At the Prairie Regional Laboratory of NRC Canada, the osmophilic yeasts used were isolated from flowers, spoiling honey and dried fruits and were mostly classified as strains of *Saccharomyces rouxii* abd *Torulopsis magnoliae* which was later reclassified as *Candida magnoliae*. Strains of the first species produced glycerol and arabitol in ratios depending on the conditions of culture and the second one glycerol and erythritol.

Peterson et al. [37] at the Forest Products Laboratory in Madison, Wisconsin studied 11 representative osmophilic yeasts of the genus *Zygosaccharomyces*. Seven strains produced high yields of polyols and four of these: *Z. nadsoni*, *Z. nussbaumeri*, *Z. richteri*, and *Z. rogus* gave yields of glycerol of 15% or more. Search for an osmophilic yeast strain that would produce glycerol as the only high-boiling constituent of the spent medium was undertaken by Hajny et al. [38]. A total of 22 cultures were screened for their glycerol-producing ability. Of these, one culture I_2B also identified as *Candida magnoliae* produced glycerol as the only polyol in any appreciable quantity. An important point made by Hajny et al. [38] and later by Button et al. [39] was that the ability to produce glycerol was lost to a great extent if the cultures were not transferred from an old lant to a new one every 3 to 4 weeks.

So far, the yeasts or yeast-like fungi which produced glycerol with high yield had a tolerance for high concentrations of sugars. An extensive study on the subject of salt tolerance of yeast was made by Onishi and his coworkers [40–44]. Their work had begun as an investigation of the organisms used in the production of soy sauce, but when sugar-tolerant yeasts were shown to produce polyols, these workers also looked for the formation of the same products by the cultures isolated from salty environments. They were able to show that these organisms, including the new species *Pichia miso* and a large number of other salt-tolerant yeasts, gave yields of polyols roughly equivalent to those produced by the sugar tolerant yeasts. A process for the simultaneous production of D-arabitol, erythritol and glycerol by *P. miso* or

Debaryomyces mogii was the basis of a patent by Onishi [40]. With *P. miso*, 28.8% of glucose utilized was converted to D-arabitol, 2.0% to erythritol and 20% to glycerol. Recently, *Pichia farinosa* has been used to produce a high yield of glycerol [45].

The essential differences in the osmophilic yeast process as compared to the steered normal yeast process are (a) that it employs aerobic rather than microaerophilic or anaerobic conditions for growth, (b) no addition of steering agents is needed (c) a considerably higher sugar content can be consumed by the osmophilic yeasts (d) an improved rate of conversion of sugar to product and much improved yield is achieved. Under optimum conditions of aeration, glycerol formation can be maximized and ethanol production reduced to a negligible amount. Two review articles on the production of polyhydroxy alcohols by osmotolerant yeasts by Spencer and his coworkers summarize the research until 1978 [46, 47]. A brief summary of the research investigations on osmophilic yeast process are up-dated here.

Formation of glycerol is an integral part of the normal growth processes of the osmophilic yeasts, however, yields of glycerol can be influenced by the conditions of growth. The composition of the medium, the level of aeration, and the temperature have the greatest effect on yields and the rate of production. The nature of the carbon source depends on the yeast species. *S. rouxii* used glucose, maltose and sometimes galactose, trehalose and sucrose and produced glycerol and arabitol during growth on these substrates. *C. magnoliae* assimilated glucose, galactose, raffinose and sucrose, but not maltose or trehalose. *P. farinosa*, however, used a wider range of common sugars including sorbose, cellobiose, xylose, ribose and sometimes lactose and L-arabinose [47]. In practice, only a few of the possible sugars were tested as carbon sources for the formation of glycerol. Glucose was the most common carbon source used for the laboratory studies [35–37, 48]. Only recently, cane molasses (containing mainly sucrose) have been used for laboratory investigation [49–51]. The concentration of sugars and salts affected the yields and the ratio of products formed. With *S. rouxii*, increasing the concentration of glucose raised the yield of glycerol without bringing about much change in arabitol yield [52]. For production of glycerol by *C. magnoliae*, yield of glycerol increased at first, as glucose concentration was raised and then decreased after an optimum concentration was passed [38]. The optimum concentration of glucose was not fixed, but was related to the aeration rate so that, at higher availability of oxygen, the optimum sugar concentration for maximum glycerol could be increased considerably. Hajny et al. [38] obtained a conversion of glucose to glycerol as high as 42.5% and a final glycerol concentration of 7.9% using *C. magnoliae* I_2B after 4 d. With the same yeast high conversion up to 43% polyol (measured as glycerol) with the final concentration of 11% was obtained after 5 d recently [51] with a molasses medium. As far as nitrogen sources go, a wide variety of them were used in the production of glycerol. The most common nitrogen source was yeast extract along with a small amount of urea or ammonium salts [35, 37, 38]. However, when molasses was used as a carbon source, the need to add yeast extract was avoided [49–51].

The effect of aeration on polyol production via osmophilic yeast is not clearly understood. In a way, an osmotolerant yeast (e.g. *S. rouxii*) like *S. cerevisiae* exhibited, a similar aeration effect i.e. at low levels of aeration, conversion of glucose (or other sugars) to ethanol was relatively high and decreased as the level of aeration was increased. However, it differed from *S. cerevisiae* in respect to sugar utilization and product formation at a high aeration rate. It had been observed that at a high rate of aeration, sugar utilization by osmophilic yeast increased or remained the same while the yield of ethanol decreased and that of glycerol increased [36–38]. This contrasted with *S. cerevisiae* where increased aeration rate decreased glucose utilization and increased the biomass product. There seemed to be an optimum level of aeration and agitation for a given composition of the medium for maximum production of glycerol via osmophilic yeasts. Since most of the investigations on osmophilic yeast process were carried out at shake flask scale, not much quantitative information was available on this aspect of the process except for a few recent papers [53–55]. The rate of oxygen supply per unit of biomass, which was the critical factor determining the percentage conversion of glucose to glycerol (and other polyols) was determined by the density of the yeast population in relation to the rate of oxygen supply to the medium and more quantitative investigations are needed to establish these relations [54].

Temperature increased the rate of glucose utilization for a small range of temperatures. Temperature sometimes also affected the ratio of product mix during the process [48, 52]. The optimum temperature for glycerol production by *C. magnoliae* and various strains of *S. rouxii* lay between 30 and 35 °C. Both the yield of glycerol and cell growth were greatly reduced at 40 °C. Similar temperature effects have been reported for *C. magnoliae* I_2B on a molasses and urea medium [51].

2.2.4 Algal Cultivation Process

The production of glycerol via algal cultivation may, presently, look like an unattractive route as it requires specific agroclimatic conditions. However, it is one of the most economical, requiring the least amount of energy input other than sunlight. It is a most promising potential route for glycerol production in the future, especially for tropical countries where sunlight is available in abundance [56]. There are many algae species which are known to produce intracellular glycerol as a survival mechanism against high concentrations of salt in the environment [57–59]. Under appropriate conditions, many species of salt-tolerant green algae *Dunaliella* and *Astermonas* produce and accumulate large amounts of intracellular glycerol, consisting of as much as 50% of the weight of the dried algae [60, 61]. Thus, large scale cultivation of the glycerol-producing algae followed by cell harvesting and product recovery steps represents an alternative process for glycerol production. The most attractive feature of this relatively new bioconversion route is the direct utilization of solar energy for the production of glycerol and as a by-product it provides animal feed and β-carotene (pro-vitamin A and a natural food coloring agent). Consequently, there are no waste treatment problems associated with this process.

Dunaliella and *Astermonas*, the algal species of interest, are found in brackish water such as the Dead Sea. Unlike most living cells, these algal cells can survive in a high salinity environment containing NaCl at concentrations greater than 4 M. With the exception of the high salinity environment, the growth requirements of this algae are fairly standard. In addition to sunlight, the algae cultivation requires a nutrient supply of CO_2, nitrate, phosphate and trace metals in a medium of pH 7–9 at moderate temperatures 10–40 °C [56]. The algae seem to adopt easily to a wide range of salt concentrations, but the growth of the algae is a strong function of the NaCl concentration. For example, algae grown at 4 M NaCl multiply at a rate which approximates only to about one-third that of algae grown under optimal conditions [60]. However, the intracellular glycerol content of the algae is directly proportional to the extracellular salt concentration and the linear relationship is maintained over a broad range of salt concentrations, from 0.5 M to 4.5 M [61].

Microscopic observations show that *Dunaliella* cells behave like perfect osmometers, rapidly shrinking or swelling under hypertonic or hypotonic conditions, respectively (Fig. 4). The absence of a rigid polysaccharide cell wall permits a rapid adjustment of the intracellular osmotic pressure by fluxes of water through the cytoplasmic membrane. Thereafter, the cells slowly return to their original ellipsoid-like shape through a phase of metabolic adjustment. During the metabolic adjustment period under hypertonic conditions, the algae produce and accumulate glycerol above the original level, while under hypotonic conditions, the algae reduces the glycerol content to below the original level. In either case, water flows through the cytoplasmic membrane in response to the new level of intracellular glycerol so that at a steady state, the original cell volume is restored. The enzymes which have been detected in *Dunaliella* and are likely to be involved in its osmoregulatory response are a subject of discussion in a separate section.

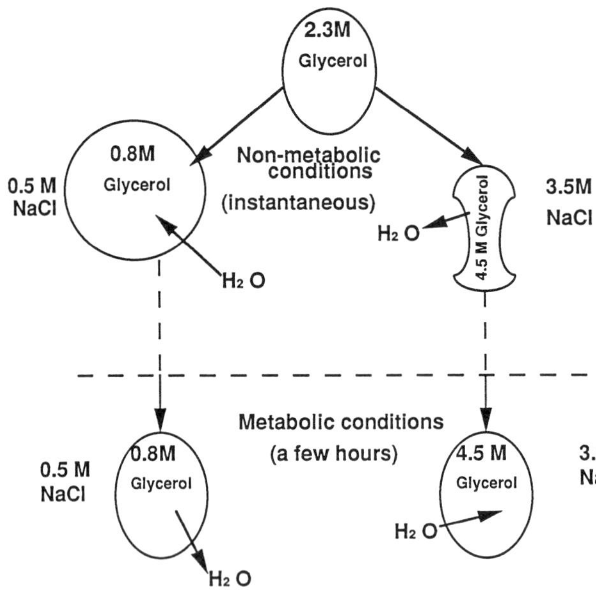

Fig. 4. Osmoregulation in halotolerant wall-less algae schematic representation of the adjustment of *Dunaliella* to hypertonic and hypotonic conditions (Ref. 61)

Based on the information on *Dunaliella* algae on a small scale, a process flow sheet for the glycerol production (Fig. 5) via algae has been proposed [56, 61, 63]. In this process, sea water and stack gas (i.e. CO_2) represented the major raw materials, while sunlight served as the major energy input. Sea water was the logical source of water

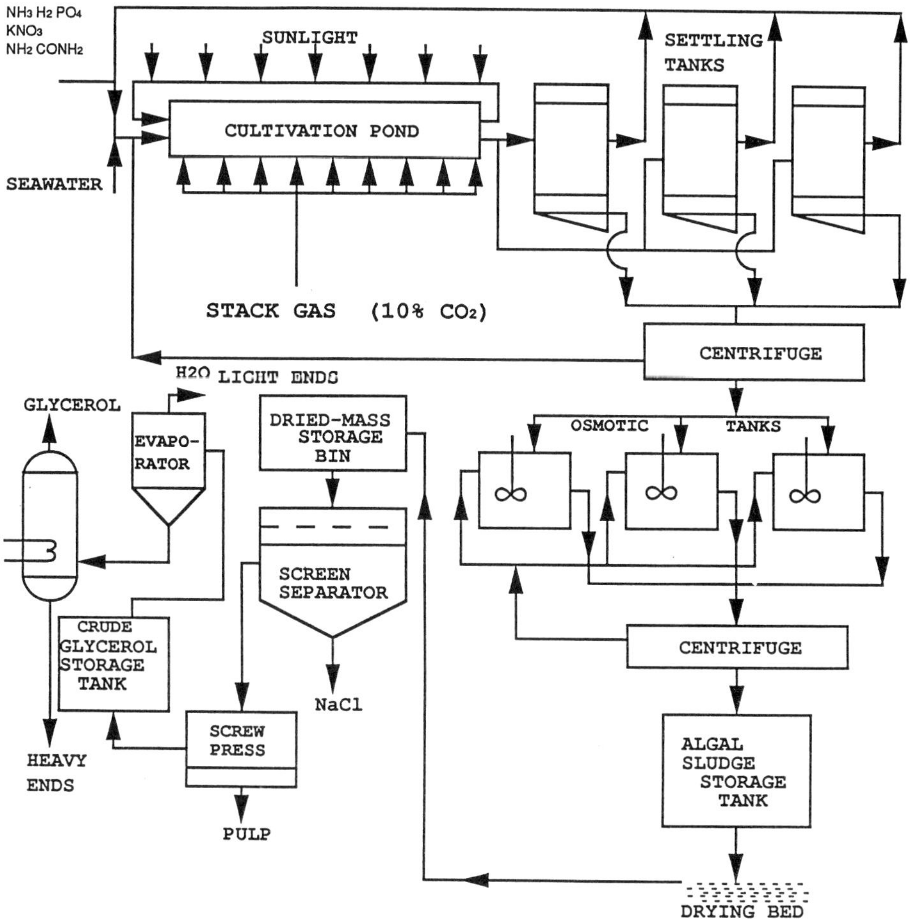

Fig. 5. Process flowsheet for glycerol production via algae cultivation (Refs. 56, 63)

as the cultivations were to be carried out in a high salinity medium. Also proposed was the use of stack gas from boiler plants of medium size manufacturing facilities. Due to the difference of optimal NaCl concentrations for the production of biomass versus the production of intracellular glycerol, the two-step process for glycerol production was suggested. For the cultivation step, NaCl concentration was controlled at 0.5 M so as to favor rapid cell growth and reproduction. The cultivation pond open to air, had a shallow depth of 20–30 cm to enable maximum exposure to solar illumination. Before the sea water was fed to the first cultivation pond, it was enriched with KNO_3, $NH_4H_2PO_4$, and NaCl to concentrations of about 4.0 mM, 1.0 mM and 0.5 M, respectively. For the removal of sulphur compounds and other toxic materials, the stack gas was scrubbed before it was distributed continuously by spargers at the bottom of the pond. The objective of the cultivation step was to produce a product stream containing a high concentration of viable algal cells. The algae grown in the cultivation tank was fed to the settling tanks and centrifuged before it was sent down to the osmotic tank containing a high concentration of salt (about 4.5 M NaCl) where glycerol was synthesized. The volume requirement of the osmotic shock tank was much less than that of cultivation tanks as the objective of the osmotic tank was the production of intracellular glycerol only. In the osmotic shock tank, the osmotic shock activated the survival mechanism of algae whereby intracellular deposits of polysaccharides were converted into glycerol with a minimum of solar energy and CO_2, and virtually no new cells were produced in this step. These first two steps could be operated batch-wise. The residence time in the cultivating tank could be 1–2 d while a few hours might be sufficient for the osmotic tank.

3 Downstream Processing

The economics of downstream processing, in the long run, are going to be decisive factors in the commercialization of the biochemical processes of glycerol (a high boiling point chemical) production, as it constitutes a major component of the overall cost. Little attention is paid to this aspect of biotechnology in general [64]. While much attention in biotechnology today has been directed at the discoveries and developments in fundamental molecular biology, molecular genetics, novel bioreactor designs and scale-up, not much advancement has been made in the downstream processing aspect to make the product recovery easier. The commercialization of new processes of biotechnology requires a coordinated coupling of unit operations in order to develop overall efficient process. The recovery of glycerol from spent medium or cultivated algae is no exception to this general rule. In this section, the major developments in the recovery of glycerol processes are highlighted in the same order as they are described earlier.

3.1 Glycerol via Chemical Processes

The spent lye resulting from current soap-making processes generally contain 8–15% glycerol; sweet waters from hydrolysis of fats contain as much as 20% glycerol [3]. Conventionally, for the separation of glycerol from soap lye/fats hydrolyzate, it is first concentrated in the multi-effect evaporator. In the case of soap lye, upon concentration, salt is formed which is to be filtered off. In the case of sweet water, the concentrated glycerol (80% or so) is separated from the small amount of sludge that forms during concentration as a result of lime treatment. In the second step, refining of glycerol is carried out by distillation followed by treatment with active carbon. For soap lye, the composition of the concentrated crude is around 80% glycerol, 7% water, 3% organic residues and 10% ash. The concentrated crude from fat hydrolyzates are generally of a better quality than the soap lye crude and has little or no salt. Distillation of glycerol under atmospheric pressure is not practicable since it polymerizes and decomposes glycerol to some extent at 20_4 °C. A combination of vacuum and steam distillation is used in which the vapors are passed from the distillation still through three stages of condensors. Final purification of glycerol is accompanied by carbon bleaching. Similar steps are also used for recovery of glycerol from the chemical synthesis process where glycerol is obtained as an aqueous solution containing about 5% glycerol and a large amount of salt [56]. It has been reported that approximately 34.6 MJ of distillation energy is consumed for the production of 1 kg of glycerol from the propylene synthesis process, making it the most energy intensive in terms of distillation among 50 important chemical and petrochemical processes [65].

The major portion of the glycerol in the USA is still refined by the methods described above. However, several refineries employ the ion-exchange system [3] in which ionized solids are high, as in soap-lye crude and synthetic glycerol from propylene, ion-exchange treatment is used to separate the ionized material from the non-ionized (mainly glycerol) [66, 67]. The ion-exchange process has the advantage over the conventional process as the distillation step which is a very energy intensive process is totally avoided in the former. Moreover, higher glycerol recovery yield might be expected in the ion-exchange process which is non-thermal (i.e. operating temperatures are low) over distillation as organic impurities and glycerol tend to polymerize at high temperatures, decreasing the recovery yield. The potential of ion-exchange resin should be realized especially in the glycerol recovery from the complex medium of the biochemical processes as discussed in the following paragraphs.

3.2 Glycerol via Yeasts and Bacterial Processes

The recovery of glycerol from any yeast processes would normally involve the following steps. First the cells need to be separated out from the spent medium and the cells recycled if required. The clarified medium may be treated with an acid or base to neutralize it and precipitate out any dissolved solids, followed by filtration. The treated medium is concentrated in the multi-effect evaporators with the consequent removal of volatiles like acetaldehyde, ethanol, etc. and purified using a solvent extraction method. Finally, it is distilled off under vacuum to obtain pure glycerol. During World War I, glycerol was recovered on a large scale in Germany and Austria from a similar sequence of steps to the ones described above. The exact details of the recovery processes are not available as most of the processes were patented. As far as is known, in the commercial production of glycerol in Germany and Austria, pure sugar was used as a substrate and the recovery yield of dynamite glycerol from the industrial processes rarely exceeded 60% and was normally less than 50% of that present in the spent medium [5]. Tremendous losses occurred in the recovery operations so that usually 10–12 kg of refined sugar was required to produce 1 kg of dynamite grade glycerol in the German practice. The recovery problem would have been greatly intensified with the use of molasses which has 30–40% non-sugar solids.

Later, several sulfite-steered yeast processes for glycerol production from molasses were developed [68] and one of the processes was used by ICI. In this process, the spent medium was first distilled for the removal of acetaldehyde and alcohol, leaving behind a glycerol solution (5–10%) which was then concentrated, solvent extracted and refined. In the recovery of glycerol from the spent medium, the solvent extraction step became a necessity owing to the large amount of soluble impurities which were present in the molasses and the sulphite added during the process. The use of solvent for glycerol recovery made the processing cost prohibitive as a large percentage (normally 5–10%) was lost in the liquid-liquid extraction.

Underkofler and Hickey [5] outlined the difficulties in glycerol recovery in the sodium sulfite-bisulfite-steered process and proposed several alternatives as described earlier to reduce the content of sulfite in the growth medium. Taking clues from this important observation, the Wisconsin group carried out a pilot plant study on continuous yeast process of high test molasses where they used ion-exchange resins to recover glycerol in high yields [11]. This was the first time that the use of ion-exchange resin from the complex medium of this process was reported. The particular method chosen for the pilot plant unit was a combination of ion-exclusion, ion-exchange and evaporation similar to that described earlier. In the pilot plant investigation, the ionic impurities were removed by over 95% while non-ionic impurities were reduced by about 92% in an ion-exchange column [11]. The product stream from the exclusion column was ion-exchanged to obtain an aqueous glycerol solution of high purity. The exchange unit consisted of two anion-cation pairs and a mixed bed. The product thus obtained upon evaporating the ion-exchanged solution was water white and it met almost all U.S.P. specifications.

Most of the published material on glycerol recovery investigations are confined to the sulfite-steered yeast processes. Though the recovery of glycerol from the osmophilic yeast process would be essentially the same as that of the sulfite-steered method. Roxburgh et al. [69] described a method in detail and a patent was also obtained on the simultaneous recovery of D-arabitol, erythritol and glycerol [40]. Essentially, the method consisted of removing yeast cells by settling or centrifuging in a primary clarification step after acidification of the medium. Most of the water was then removed by evaporation, preferably under a vacuum. Hot ethanol was then added, and the gums were filtered off and the solution was decolorized with activated charcoal. Erythritol and arabitol are only slightly soluble in cold mixtures of glycerol and ethanol, and they crystallized from the solutions upon cooling. After the solids were removed by filtration, alcohol was recovered from the liquid by distillation at atmospheric pressure, and glycerol was vacuum distilled in the conventional manner and recovered in fair yield. Since the process involved the use of a solvent as well as a distillation step, the process would not be cost effective. The recovery process might be simplified considerably if the spent medium contained predominantly glycerol. In such a case, the ion-exchange resins could possible be used for the recovery of glycerol in pure form without having to carry out the distillation.

In this process, glycerol separation from the other high boiling point chemicals, 2,3-butanediol and lactic acid would be quite complicated and nothing is reported in the literature on the separation of glycerol from 2,3-butanediol and lactic acid.

3.3 Glycerol via Algal Cultivation

The process of glycerol production from algae as described here is based on a proposed scheme by Chen and Chi where actual data are lacking [56]. It might, perhaps, be premature to write about the recovery scheme which is not supported by experimental evidence. However, a logical sequence of downstream processing steps for algae is given in the flow diagram (Fig. 5). According to this scheme the stream leaving the osmotic tank is further concentrated by centrifuging and the algal sludge thus obtained is sun dried by spreading it into a thin layer of 0.25–0.5 cm thick. A large amount of both intracellular and extracellular water is evaporated by direct solar illumination. As a result, the cells form chiplike aggregates coated with crystalline salts which precipitate during water evaporation. The mixture of algal chips and crystalline salts is transported to a screen separator where the salts are removed and recycled. The algal chips are then fed to a continuous screw press where the cells are disrupted and the intracellular fluid literally squeezed out. The glycerol-rich liquid stream leaving the press is first fed to an evaporator before it is distilled off under vacuum. All of the unit operations following the osmotic shock tank could be in either mode (continuous or batch). Pilot plant work on glycerol production from algae was carried out by Koor Food Ltd. in Israel and they concluded that glycerol production in the near future was not profitable due to the expensive harvesting step [70]. Again, no data is available to comment upon the status of research in this area. Western Biotechnology Ltd. in Perth, Australia cultivated *Dunaliella* on a large scale, but their interest was confined to the recovery of β-carotene from algae biomass [71]. Based on the preliminary study on algae cultivation and harvesting, it can be safely said that until new methods of harvesting are found where centrifuging steps may be done away with completely it would be quite an energy-intensive step [63].

From the limited literature, it is imperative that more research investigations are carried out to simplify the recovery processes and an earnest effort is needed to make the recovery process less energy intensive. With the recent developments in membrane technology for reverse osmosis and ultrafiltration, it may be possible to concentrate and purify glycerol directly from the dilute glycerol solutions of yeast processes. Murakaml and Igarashi [72] reported that with the use of suitable synthetic membranes for reverse osmosis, glycerol solutions could be concentrated to 97%. Desai et al. [73] successfully made use of reverse osmosis with cellulose acetate butyrate membranes for the recovery of glycerol from a petrochemical effluent stream. Some preliminary investigations on the concentration of glycerol by reverse osmosis has also been carried out by Vijaikishore and Karanth recently [74]. More investigations are needed to apply membrane technology to the recovery of glycerol from the dilute solutions. Investigations are also needed to find cheaper ways of harvesting large volumes of liquid in algae cultivation.

4 Biosynthesis of Glycerol

The discussion of glycerol production via biochemical routes would remain incomplete if metabolic pathways inside the cell for the biosynthesis of glycerol in various processes were not studied in detail. The study of metabolic pathways is as important in biotechnology as the study of catalysis in chemical technology and it is useful for the understanding of enzymatic reactions, possible intermediates and by-products, effect of operating parameters like pH, dissolved oxygen supply etc. on the overall rate of biosynthesis. It would be completely out of place in this section to cover what is known of the very diverse metabolic activities of microorganisms. However, an attempt should be made to highlight the landmarks in the developments of metabolic pathways of the organisms and the need to carry on the research in the area where little is known. The references can be consulted for the detailed discussion on this subject.

4.1 Normal Yeast Metabolism

Normal yeast is one of the few organisms whose metabolism has been investigated in great detail. It is briefly summarized here in relation to glycerol production. Figure 6 gives all the steps of the glycolytic pathway which is also known as the Embden-Meyerhof-Parnas (EMP) pathway, including all the enzymes involved. All of these enzymes are generally found free in yeast cell cytoplasma and most of them have been isolated in a pure form and their enzymatic, physical, chemical properties properly characteriz-

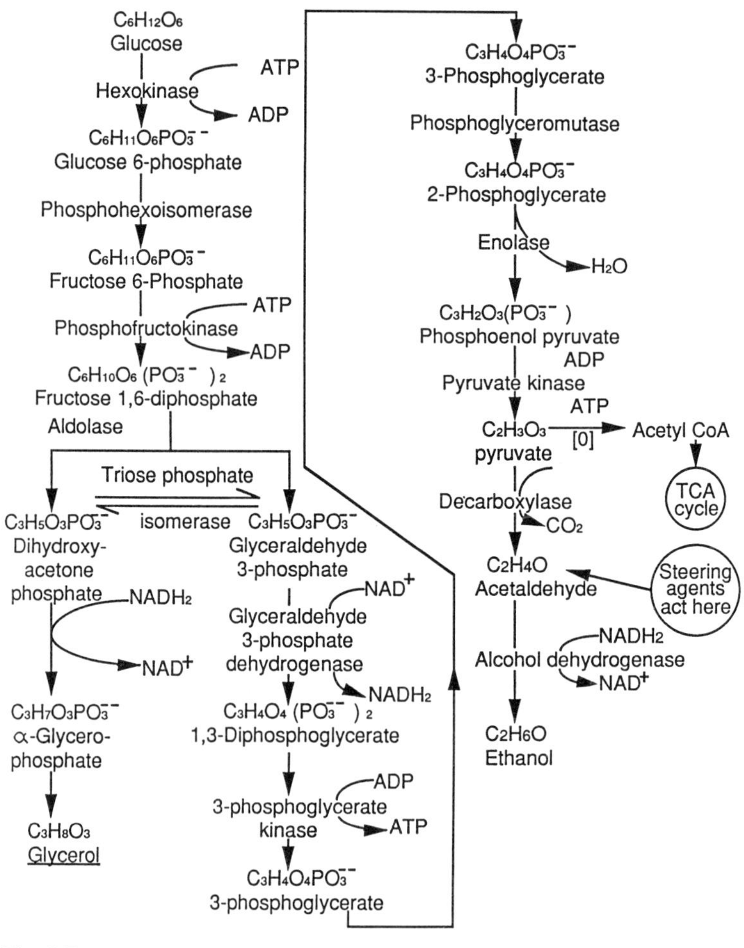

Fig. 6. Formation of glycerol and ethanol by anaerobic dissimilation of glucose via EMP pathway. For aerobic process TCA cycle is incorporated

ed. The interesting and important aspect of glycerol formation is the mechanism which controls the relative yields of glycerol and ethanol. Formation of glycerol results from reduction and dephosphorylation of dihydroxyacetone phosphate, formed by scission of fructose 1,6-diphosphate by aldolase. Under anaerobic conditions of growth of *S. cerevisiae*, most of the glucose is converted to ethanol. During the process, NAD^+ is first reduced to NADH, then reoxidized during reduction of acetaldehyde in the final stage of ethanol formation. A little of the NADH is diverted and used up in the reduction of dihydroxy acetone phosphate to α-glycero-phosphate, which is then dephosphorylated, yielding glycerol. The objective for the production of glycerol by yeast glycolysis is to influence the reactions of the EMP pathway in such a way as to suppress formation of ethanol and divert more of the NADH formed earlier to reduction of dihydroxy acetone phosphate to α-glycero-phosphate.

In the steered yeast processes, this is achieved by the use of sulfites, bisulfites or alkalies. Acetaldehyde binds to bisulfate to form a complex which cannot react with NADH and this in turn reacts with dihydroxyacetone phosphate, yielding glycerol. Once acetaldehyde is fixed by sulfite, the stoichiometric relationship as expressed by Eq. (2) is followed. From the stoichiometry, a 51% theoretical yield of glycerol (based on the hexose consumed) is possible, however, the theoretical yield has never been realized and major products of the process are ethanol in addition to glycerol and acetaldehyde. It means that part of the glucose undergoes dissimilation into ethanol and CO_2 by the following stoichiometry:

$$C_6H_{12}O_6 \rightarrow 2\,C_2H_5OH + 2\,CO_2 \qquad (5)$$

The control mechanism in the alkaline-steered yeast process depends on the occurrence of a Canizzaro reaction in which one molecule of acetaldehyde is oxidized to acetate, while a second is simultaneously reduced to ethanol. For sometime, this reaction was assumed to be nonenzymatic, but it does not occur in the absence of living yeast cells and so represents an alternation induced by alkali in the normal metabolism [17]. Thus, one molecule of acetaldehyde functions as a hydrogen donor, resulting in the removal of another molecule of acetaldehyde from the normal reaction system, so that again there is a surplus of NADH, which can take part in the reduction of dihydroxyacetone phosphate, leading to increased yields of glycerol. If all of the acetaldehyde produced by glycolysis appears in a Canizzaro reaction, the stoichiometry would be as given by Eq. (4). Again, the theoretical yield of glycerol based on hexose consumption would be 51%, however, this is never achieved in practice. Moreover, a much higher yield of alcohol is observed in any alkali yeast process while glycerol yield is limited to about 25% which implies that part of glucose dissimilation is through Eq. (5).

4.2 Osmotolerant Yeast Metabolism

Steering agents are not necessary in processes for glycerol production using osmotolerant yeasts, and the controlling factors are not known, except that they appear to be related to the respiratory metabolism. An adequate, but not excessive, supply of oxygen is necessary for maximum yields of products and the growth of the organism. The conditions of suboptimum aeration results in increased quantities of ethanol formation and a considerable decrease in the yield of glycerol. The basic differences in the metabolic behavior of osmophilic yeast from normal yeast is discussed here. The glycerol production metabolism via osmotolerant yeast is given in detail by Spencer [46] and Spencer and Spencer [47]. But here the main points are highlighted with the intention of arousing curiosity about the metabolism of these yeasts and showing the scope for further investigation. It is well known that *S. cerevisiae* under aerobic conditions produced cell biomass with a concurrent decrease in the rate of glucose utilization which was generally referred to as the 'Pasteur effect'. Sols et al. [75] suggested a mechanism for *S. cerevisiae* in which regulation of the balance between glycolytic pathways depended upon allosteric feedback mechanisms affecting several enzymes of the EMP pathway and the first irreversible step of the TCA cycle. The scheme had many attractive features. However, it was not studied in the osmotolerant yeasts, and it did not seem to account for the concomitant increase in production of glycerol and other polyols with increasing aeration while decreasing ethanol formation. It is, therefore, expected that investigation of oxygen uptake rate in osmotolerant yeast might yield more insight than has so far been obtained using *S. cerevisiae*. Osmotolerant yeasts have been shown to contain numerous polyol dehydrogenases in addition to glycerol dehydrogenase which are necessary in the final stages of formation of the various polyhydroxy alcohol along with glycerol. Polyol dehydrogenases catalyze the last reaction or reactions on the pathways leading to formation of the different polyols, and they usually reduce one or the other of the corresponding ketoses. Some of the dehydrogenases which oxidize mannitol, sorbitol, dulcitol, erythritol, and arabitol have been identified [76, 77]. Since formation of polyols (in addition to glycerol) apparently proceeds via a number of phosphorylated intermediates, a variety of phosphatases are necessary for the final step of the process, leading to conversion of polyol phosphates to the corresponding polyols and these phosphatases have been detected in some osmotolerant yeasts like *S. mellis* [78] and *S. rouxii* [79].

The identification of additional enzymes in osmotolerant yeasts does not elucidate the mechanism of glycerol formation in these cells. Some evidence concerning metabolic pathways of the formation of glycerol, ethanol and other polyols had been obtained with glucose labelling techniques in a strain of osmotolerant yeast *S. rouxii* [80]. Glucose-1-^{14}C and glucose-3-^{14}C were used for the study and they found that the distribution of ^{14}C in ethanol and glycerol agreed with the assumption that they were formed mainly from triose phosphate produced via the EMP pathway (Fig. 6). However, the role played by oxygen in the osmotolerant yeast cannot be explained merely by the EMP pathway which is essentially a glycolytic pathway. How the regulatory control exercised in osmotolerant yeast in the presence and absence of oxygen is different from that of normal yeast remains unanswered. Despite the evidence in favor of the EMP pathway, the factors which influence the formation of high yields of glycerol and other polyhydroxy alcohols in the osmophilic yeast process has not progressed past the realm of speculation.

On the basis of carbon balance carried out by Button et al. [39] on glycerol production via *C. magnoliae* I_2B where about 99% of the carbon was accounted for, the products carbon dioxide and glycerol were produced in a 3:1 molar ratio while glucose was converted to glycerol on an equal molar basis and the following stoichiometry was proposed:

$$2\,C_6H_{12}O_6 + 5\,O_2 \rightarrow 2\,C_3H_8O_3 + 6\,CO_2 + 4\,H_2O \tag{6}$$

The ratio of carbon dioxide to oxygen in this equation is 1:1.2 which compared well with the measured respiratory quotient of 1.17. The theoretical yield as per the stoichiometry is about 51% of the hexose consumed. Experimental evidence also points to the fact that the maximum yield obtained is less than 51%. However, what remains unanswered is the exact role of O_2 in the metabolic pathways except that it relates to the TCA cycle with consequent regeneration of NAD^+. Is the proposed stoichiometric Eq. (6) explainable by the metabolic pathways? Is it possible to change the stoichiometry to some other form by changing the experimental conditions so that a higher glycerol yield is realized?

Another important aspect which attracted the attention of many investigators was related to the nature of osmotolerance in yeasts. In the case of salt tolerance yeasts, Onishi [44] contributed most significantly while Brown [81] made some progress in the understanding of the nature of sugar tolerant yeasts. Onishi [44] worked on hypotheses for the mechanism of tolerance towards high concentrations of salt or sugar: either the solutes (sugar or salt) did not penetrate the cell in high concentration, or if they did the enzyme system of the cell was specially resistant to inactivation by them. In the absence of any special halotolerant enzyme system, the latter hypothesis was discarded and it was suggested that osmotolerance in yeasts depended on some kind of exclusion mechanism which keeps the intracellular concentrations of sugar and salt low.

In regard to sugar tolerance of osmophilic yeasts, Brown [81] confirmed that the enzymes were fundamentally similar in both tolerant and non-tolerant organisms in line with Onishi's [44] observations. But intracellular polyol concentrations were different in such a way that the impact of the environment on the intracellular physiology of tolerant species was lessened. This suggested mechanism for osmophilic yeast was similar to osmoregulatory mechanism described later in this section for algae.

4.3 Bacterial Metabolism

Since the production of glycerol by bacterial process was realized during the investigation of 2,3-butanediol producing organisms, it is only natural to find the mechanism for its production in the context of 2,3-butanediol. It was observed that pyruvic acid was readily converted by 2,3-butanediol producing bacteria to acetoin, an important intermediate, which in turn was reduced to give 2,3-butanediol. The synthesis of acetoin was closely linked to the decarboxylation of pyruvic acid [29]. 2,3-Butanediol producing bacteria were also known to follow the EMP pathway for glucose catabolism [82]. As per this pathway, glucose is split into two 3-carbon chain compounds, yielding dihydroxyacetone phosphate and glyceraldehyde phosphate similar to yeast glycolysis. The former is converted to glycerol following the reactions shown for the yeast metabolism earlier, while the latter gives rise to the pyruvic acid

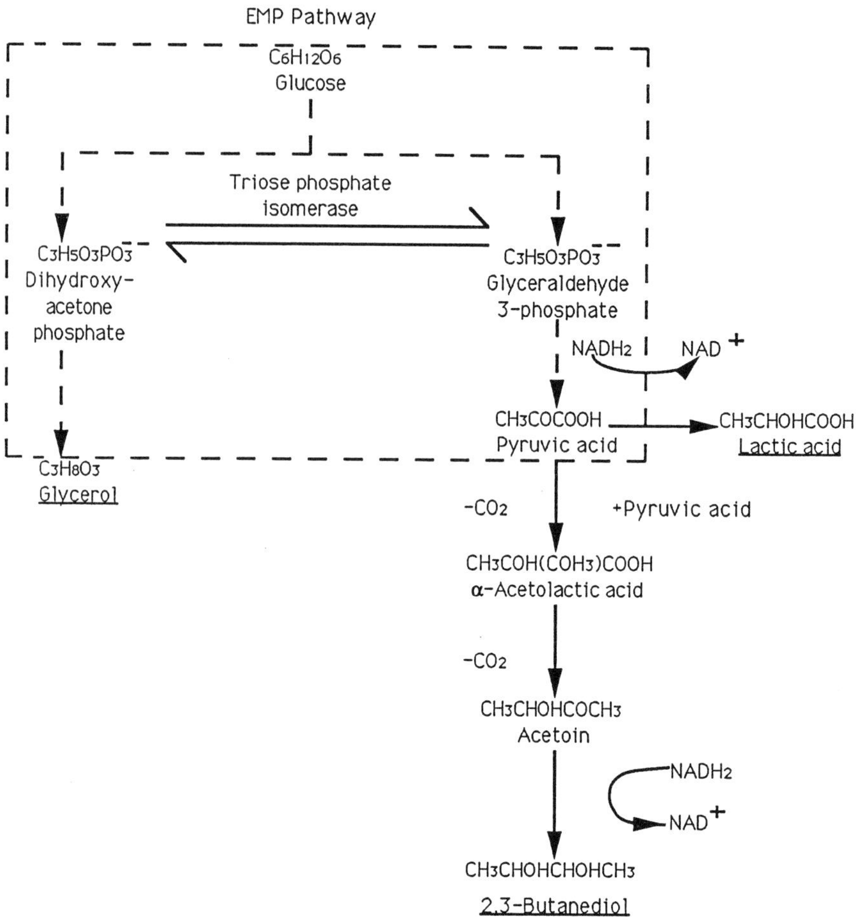

Fig. 7. Formation of glycerol, 2,3-butanediol and lactic acid by *Bacillus subtilis* (Ford's strain)

which is eventually converted to 2,3-butanediol as per the scheme given in Fig. 7 but it does not give the complete EMP pathway, the details are given in Fig. 6. The pyruvic acid produced by the EMP pathway combined with another pyruvic acid in the presence of carboxylase enzyme to form α-acetolactic acid which is again decarboxylated into acetoin. Acetoin is finally reduced to 2,3-butanediol by the NAD-NADH coupled enzymatic reaction.

It was reported that some of the strains of *B. subtilis* converted as much as 86% of the glucose by the above mechanism. This mechanism would give rise to the stoichiometry Eq. (4) if it is assumed that no lactic acid is formed. Another major product obtained from the bacterial process was lactic acid which would be produced if part pyruvic acid was directly reduced by the NAD-NADH coupled enzymatic reaction in parallel. According to this mechanism, the maximum theoretical glycerol yield is only 34% which does not compare well with the theoretical yield of 51% in yeast and osmophilic yeast processes.

4.4 Osmoregulatory Mechanism of Algae

Microscopic observations on *Dunaliella* cells as perfect osmometers, rapidly shrinking or swelling under hypertonic or hypotonic conditions was described earlier (Fig. 4). The kinetics and the mechanism of synthesis and elimination of intracellular algal glycerol upon transition from low to high concentration of salt or vice versa was thoroughly investigated, the summary of these observations is given here [59, 83–85]. The glycerol synthesis or elimination within the algal cell was very rapid and detected within minutes after the transition in the external environment of algae.

Three unique enzymes had been detected in *Dunaliella* and *A. gracilis* which were thought likely to be involved in its osmoregulatory response. These enzymes were: (a) $NADP^+$ dependent dihydroxy acetone (DHA) reductase which catalyzed the interconversion of DHA and glycerol (b) DHA kinase which catalyzed the phosphorylation of DHA (c) Glycerol-1-phosphatase which dephosphorylated α-glycero-

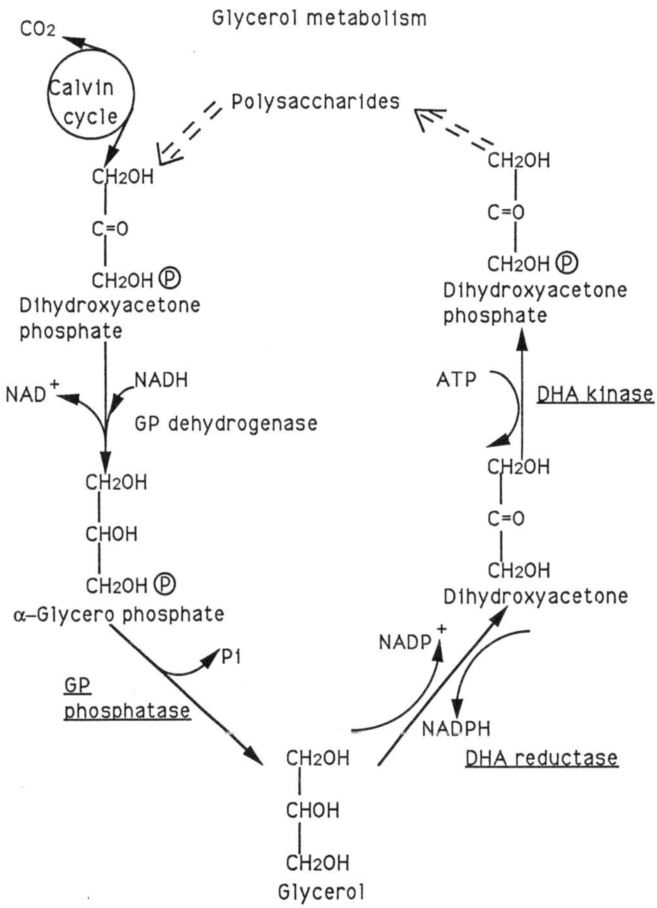

Fig. 8. A hypothetical metabolic pathway of glycerol biosynthesis and degradation in *Dunaliella* (Ref. 61)

phosphate. Taking into consideration the presence of these enzymes in *Dunaliella* and *A. gracilis*, a reasonable hypothetical scheme of the osmoregulatory metabolism of these algae was proposed [61] as shown in Fig. 8. Glycerol accumulated by production of triose phosphate via photosynthesis or from polysaccharide degradation followed by reduction to α-glycero-phosphate and dephosphorylation. Conversion of glycerol back to a polysaccharide may proceed via oxidation to dihydroxyacetone and phosphorylation to dihydroxy-acetone phosphate. The enzymatic system as described seems to work very efficiently as a mechanism to absorb shocks given to the algae environmental conditions. At normal temperatures of 60 °C or so, glycerol may be released by algae to the environment. In contrast osmophilic yeasts are also known to produce glycerol as a survival mechanism against osmotic shock, but they continuously release glycerol to the medium.

5 Comparative Study of the Processes

Six main processes to produce glycerol have been briefly described. Out of these two chemical processes, namely, (a) by-product of soap/fatty acid industries, and (b) chemical synthesis from propylene are currently used for commercial production. Any of the four known biochemical processes for glycerol production has the potential for commercial exploitation in the not too distant future. These processes are compared and evaluated on the basis of the following criteria:

1. raw material,
2. energy consumption,
3. productivity, yield and final concentration.

5.1 Raw Material

With the realization of the depleting petroleum reserves, it is high time that the potential for biochemical processes is now fully understood. The renewable nature of the raw materials used for the biochemical routes is the bedrock of biotechnology today and glycerol production is no exception. The supply of glycerol as a by-product of the soap and fatty acid industry is also from renewable raw materials like vegetable oil, tallow, etc. However, there is a gradual replacement of old fashioned soaps by detergents in developed countries [4] and the great shortage of edible oil in developing countries. These factors over the years have decreased the availability of glycerol from the soap and fatty acid industries as a by-product. In such a scenario, the alternate routes, based on renewable raw materials would become attractive. In this context, it may be noted that sugars (e.g. glucose, sucrose, xylose, etc.) being the most important substrates in the biochemical processes and any reduction in the manufacturing cost of sugars is likely to enhance the viability of biochemical processes based on these substrates. It may also be mentioned that enormous quantities of renewable biomass resources are available in the world for production of these sugars. In order to increase the viability of biochemical routes of glycerol production some efficient low cost process for the manufacture of the sugars must be developed. In developing countries like India, Brazil, Cuba, etc. where sugar is produced in large quantities, the by-product molasses is available in abundance, is cheap, and may be used for the yeast or bacterial processes.

Semi-arid areas and wasteland near the sea coast can provice a location for algae cultivation. The raw materials like sea water, CO_2 required for algae cultivation are easily available at no cost. Moreover, labor is also inexpensive in developing countries. This is the reason why algae cultivation has become an important route for producing animal feed in developing countries [86, 87]. Similarly *Dunaliella* and *A. gracilis* cultivation can be carried out to provide glycerol, β-carotene and animal feed simultaneously.

5.2 Energy Consumption

In chemical process industries, the processing of chemicals is generally done under extreme conditions of temperature and pressure, while biochemical processes are carried out at near room temperature and atmospheric pressure. The need to sterilize the medium neutralizes some of the biochemical route advantage, but the overall energy analysis may still favor biochemical processes. Chen and Chi [56] compared energy consumption in petrochemical and algae processes and showed that 36.4 MJ kg^{-1} glycerol was required for the petrochemical process and only 9.4 MJ kg^{-1} glycerol was sufficient for the algal process. The value of 9.4 MJ kg^{-1} glycerol appears to be an optimistic estimate, as it neglected altogether the need to centrifuge the large volume of liquid to harvest the algae and underestimated the distillation requirement. Based on the data of *D. salina*, the preliminary energy consumption estimates for glycerol production were made which showed that the energy requirement for the algal process could be as high as 36 MJ kg^{-1} glycerol [63]. This observation was corroborated by the fact that Koor's research station in Eliat, Israel abandoned glycerol production research due to the expensive harvesting of algae by centrifugation [70]. So until a better and energy-economical method of algal cultivation and harvesting are found, the algae process may not be an attractive route for glycerol production, even though the raw material cost of the process is almost nil.

Soap lye, which is a by-product of the soap industry, has to be processed to recover glycerol. The energy requirement for processing soap lye to recover glycerol in its pure form was reported to be of the order of 27 MJ kg^{-1} glycerol [56]. This energy was used to concentrate soap lye containing 6–8% glycerol to 80% crude glycerol in a multi-effect evaporators followed by distillation. The calculations can also be made for energy requirements for processing the complex medium of yeast processes by making two assumptions that glycerol concentration in the medium is of the same order of magnitude as that of soap lye and the method of recovery as applicable to soap may also be employed for glycerol recovery from the medium. The implies that the energy consumption for processing the medium of yeast process would be 27 MJ kg^{-1} glycerol, the same as that of soap lye processing. The additional energy for the yeast process would be in the form of continuous agitation and aeration of the bioreactor in the case of osmophilic yeast and agitation and CO_2 sparging/maintaining the vacuum in the case of the sulfite-steered yeast process for 5 d. This energy requirement may be in the range of 5–10 MJ kg^{-1} glycerol depending on the bioreactor parameters [88]. For bacterial process, the glycerol build up is low (approx. 1–2%), hence a very large amount of the energy would be required to concentrate and recover the glycerol. Therefore, no calculations are made for energy consumption for the bacterial process. Table 5 compares the energy consumption of the various processes. Taking the overall view of the processing energy requirements, it is apparent that no process has the definite advantage over others. Algal processes may be made the most energy economical if better techniques like autoflocculation of algal biomass is developed for *Dunaliella* and *Astermonas* species whereby avoiding the use of centrifugation altogether. The last column of Table 5 lists the energy yield of the processes defined as the ratio of total energy output to total energy input. The energy yield of the algal process could be as high as 2 while for all other processes it stands at low values of 0.29 to 0.7. The energy yield of over one implies that there is a net energy production in the process and less than one implies that there is a net energy consumption. In all of the processes except the algal process, there is a net consumption of energy. If a non-thermal method of glycerol separation and recovery like reverse osmosis, ultrafiltration, ion exclusion techniques, etc. are investigated vigorously, then glycerol production can, perhaps, be made energy economical.

Table 5. The energy requirements by various routes for glycerol production

Energy	Energy input		Energy output	Energy yield
Process	Heat value of raw material (MJ kg^{-1} of glycerol)	Processing energy required	Heat value of product	Output/input
Petrochemical	23.2[a]	36.2	17.9	0.30
Soap lye	0.0[b]	27.0	17.9	0.66
Yeast process	27.3 or 0.0[c]	32–37[d]	17.9	0.29–0.56
Algal cultivation	0.0	9–36[e]	17.9	0.5–2.0

[a] Heat value of 0.51 kg of propylene;
[b] Since soap lye is a by-product of soap manufacture, its heat value is taken to be zero;
[c] Heat value of 2.5 kg sugar less the heat of evaporation of 10% sugar solution. If molasses is used the heat value is assumed to be zero;
[d] Mixing and refining energy;
[e] This variation arises due to the uncertainties in processing of algae.
Adapted from: Ref. 56

5.3 Productivity, Yield, and Final Concentration

Productivity, a common term in biotechnology, is synonymous with reaction rate in the chemical engineering lexicon. Productivity determines the volume of a bioreactor and its measure indicates the viability of a process. Table 6 lists the productivity of all the biochemical routes for comparison. The glycerol productivities vary from a very

Table 6. Productivities, yield and final concentration of glycerol by various routes

S. No.	Biochemical route specifications and carbon sources	Mode and scale of operation		Average productivities of glycerol	Yield of glycerol	Final glycerol	Comments	Ref.
		Mode	*Scale*	($g\ l^{-1}\ d^{-1}$)	($g\ g^{-1}$)	($g\ l^{-1}$)		
1. *Sulfite/alkali steered yeast processes*:								
a.	Sulfite using blackstrap molasses	batch	12 l	11.5	0.25	55	acetaldehyde and ethanol as by-products	[8]
b.	Insoluble sulfite using dextrose	batch	2.5 l	11.6	0.23	35	acetaldehyde and ethanol as by-products	[4]
c.	Sulfite-bisulfite using dextrose	continuous	pilot plant	33.0	0.28	50	acetaldehyde and ethanol as by-products	[10]
d.	Ammonium sulfite using dextrose	batch	<1 l	—	0.20	30	patent, so available information scanty	[11]
e.	Sulfite under continuous vacuum/CO_2 sparging using molasses	i. batch	3 l	15.0	0.25	110	acetaldehyde and ethanol lost	[18, 19]
		ii. fed batch	—do—	30.0	0.25	80		
f.	Alkali steered	batch	12 l	9.0	0.23	45	ethanol and lactic acid as by-products	[16]
2. *Bacterial process*:								
	Bacillus subtilis using glucose	batch	shake flask	2	0.29	14.7	2,3-butanediol and lactic acid as by-products	[27]
3. *Osmotolerant yeast processes*:								
a.	*C. magnoliae* I_2**B** (earlier known as *T. magnoliae*) using glucose	batch	shake flask	20	0.43	79	no other by-product	[36]
b.	*C. magnoliae* I_2**B** using molasses	batch	6 l	22	0.43	110*		[49]
c.	*C. magnoliae* I_2**B** using glucose	fed batch	60 l	17	—	170		[37]
d.	*P. farinosa* using glucose	fed batch	1 l	75	—	300		[89]
4. *Algal cultivation process*:								
	Dunaliella cultivation	batch	—	0.12	0.4^{+}	0.12	β-carotene and high protein algal meal as by-products	[58]

* Total polyol measured as glycerol
$^{+}$ g glycerol per g dry algal weight.

low 0.12 $g\,l^{-1}\,d^{-1}$ for the algal process to a high 75 $g\,l^{-1}\,d^{-1}$ for the osmophilic yeast process. Low values of productivity imply that a very large sized bioreactor is required to produce glycerol as the rate of biosynthesis is slow.

Among the yeast and bacterial processes, there is a large variation in the productivity values. The productivity of 2 $g\,l^{-1}\,d^{-1}$ based on shake flask studies for the bacterial process is quite low which can, probably, be increased a few fold in the properly instrumented and controlled bioreactor. The shake flask experiments are not dependable and the scaled up investigations are needed to evaluate the viability of some of these processes. For yeast (osmophilic and nonosmophilic) processes, the productivities are of the same order of magnitude and the variation among them only indicate that a lot of improvement is possible using different modes of operation (e.g. batch, continuous or fed batch).

The productivities for glycerol production may be compared with the other well known biochemical process like bioalcohol (i.e. ethyl alcohol) production. One finds orders of magnitude difference between them. The maximum productivity reported for the alcohol production by *S. cerevisiae* is about 2400 $g\,l^{-1}\,d^{-1}$ which is two orders of magnitude more than that of the maximum glycerol productivity yet reported [90]. These figures reflect that more investigations are needed to improve the productivity by choosing different modes of operation and bioreactor design.

The glycerol yield and its final concentration in the reaction mixture as listed in Table 6 are two other important quantities in evaluating the processes. The product yield determines the raw material requirement and the product concentration, the process energy requirement. The higher the product yield and the concentration, the lower will be the need for raw materials and energy for separation respectively. On both of these counts, the yeasts processes compare well with the bioalcohol process where the alcohol yield and its concentration reported are 0.48 g ethanol per g glucose and 120 $g\,l^{-1}$ respectively [90].

Two recent reviews on the need and the prospect of producing bulk chemicals via biochemical routes by Ghose [91] and Wang [90] require special mention. The former discusses the point of view valid for developing countries while the latter highlights the economics for the developed countries. The case is made for the production of many bulk chemicals and glycerol finds special mention in both the reviews.

6 Summary

Glycerol is an important chemical whose production history has been quite turbulent and interminably linked with the World Wars I and II. During war time, its major use was in the making of explosives for the war machine. However, in recent years, its uses have been quite diverse and as an explosive, it is hardly used anymore.

Presently, it is commercially produced by two chemical routes, namely, (a) as a by-product of the soap and fatty acid industries, (b) synthesis from propylene. In developing countries, it is produced mainly as a by-product of the soap and fatty acid industries, however, developed countries use both methods for its production. With the popularization of detergent over old fashioned soap, the availability of glycerol as a by-product of the soap and fatty acid industries has been decreasing.

The realization that the raw material for the synthetic route is of a non-renewable nature, the production of glycerol by alternative routes which are considered as potential routes have been reviewed and put in proper perspective. Some of these processes can be commercially exploited in the not too distant future.

Among the biochemical routes, the literature on sulfite-/alkali-steered yeasts processes is voluminous and it dates back to the turn of the century. However, research on these processes in recent years with a view to reduce the sulfite addition is quite interesting and needs further investigation. However, with the advent of osmophilic yeast, the need to use steering agents is completely avoided and the higher glycerol yield and concentration is possible. If the recent publications are any indication, the research on it has picked up and there is a tremendous scope for further investigation on the osmophilic yeast process. The present knowledge on the production of glycerol via bacterial processes is meagre and needs to be complemented with more investigations. The bacterial process for glycerol production does not appear to hold much promise according to the available data. Algae cultivation for glycerol production has a bright future if the harvesting of algae biomass by an energy-economical method can be carried out. The present method of algae separation via centrifugation is much too expensive to be considered economical for commercialization.

As far as metabolic pathways for biosynthesis of glycerol in the cells are concerned, normal yeast has been satisfactorily described by the well known EMP pathway. It can explain the products and by-products of yeast metabolism as well as the regulatory and control aspect especially the aeration effect. However, the aeration and agitation effect is not completely resolved for osmophilic yeast metabolism by the EMP pathway and the TCA cycle alone. The role of oxygen in the osmophilic yeast process is not clear and needs more investigations. Bacterial cell metabolism seems quite reasonable with respect to the production of 2,3-butanediol and lactic acid as coproducts along with the main product glycerol. This metabolism may be studied further in order to improve the concentration of glycerol in the final medium to make it a viable process. The algal cell metabolism is well understood with respect to its role for osmoregulation in the cell. With the knowledge of osmoregulatory metabolism of algae, it has been possible to increase the intracellular glycerol yield to 0.4 g glycerol per g cell dry weight.

One of the major problems facing the production of bulk chemicals via biochemical routes has been the downstream processing and glycerol production is no exception. Very little published literature is available on this important aspect. The conventional separation methods like centrifugation, evaporation, distillation, liquid-liquid extraction, etc. do not hold much promise as they are either energy intensive or the recovery yield of the product is low. Therefore, the need to look for new methods such as ultrafiltration, reverse osmosis, ion-exclusion techniques, etc. is obvious and may provide the better alternative for the recovery of glycerol.

Comparative data of the various processes of glycerol production in terms of raw material, energy input, glycerol productivity, its yield and the final concentration are given to highlight the prospects of biochemical routes and their potential for commercialization.

7 Acknowledgements

The opportunity provided by IIT, Delhi, India and BPEC, MIT, Cambridge, USA to complete this chapter in its final form is gratefully acknowledged. The typing of the first draft by Mr. R. N. Shukla and the preliminary drawings made by Mr. S. K. Khurana is appreciated. Without the help provided by Ms. K. Comer in typing the final draft and making the drawings, the manuscript would have taken a lot longer.

8 References

1. Miner CS, Dalton NN (eds) (1953) Glycerol, Rheinhold, New York
2. Prescott SC, Dunn CG (1959) Industrial microbiology (3rd edn), McGraw Hill, New York
3. Kirk RE, Othmer D (eds) (1980), Glycerol, Encyclopedia of chemical technology, vol II, 3rd edn, John Wiley, New York, p 921
4. Austin GT (1984) Shreve's chemical process industries, 5th edn, McGraw Hill, New York, p 529
5. Underkofler LA, Hickey RJ (eds) (1954) Industrial fermentations, vol 1, Chemical Publishing Company, p 252
6. Connstein W, Lüdecke K (1921) US Patent 1,368,023
7. Connstein W, Lüdecke K (1924) US Patent 1,511,754
8. Cocking AT, Lilly CM (1922) US Patent 1,425,838
9. Freeman GG, Donald GMS (1957) Appl. Microbiol. 5: 197
10. Freeman GG, Donald GMS (1957) Appl. Microbiol. 5: 211
11. Harris JF, Hajny GJ (1960) J. Biochem. Microbiol. Technol. Eng. 2: 9
12. Fulmer EI, Underkofler LA, Hickey RJ (1945) US Patent 2,338,840
13. Fulmer EI, Underkofler LA, Hickey RJ (1947) US Patent 2,416,745
14. Eoff JR (1918) US Patent 1,288,398
15. Schade AL (1947) US Patent 2,428,766
16. Schade AL, Färber E (1947) US Patent 2,414,838
17. Freeman GG, Donald GMS Appl. Microbiol. 5: 216
18. Rahalkar AK, Sen BP (1977) Chemical Industry Development C15-18
19. Kalle GP, Naik SC, Lashkari BZ (1985) J. Fermen. Technol. 63: 231
20. Kalle GP, Naik SC (1985) J. Fermen. Technol. 63: 411
21. Wright RE, Hendershot WF, Peterson WH (1957) Appl. Microbiol. 6: 272
22. Johansson M, Sjöström JE (1984) Biotechnol. Letters 6: 49
23. Bisping B, Rehm HJ (1982) European J. Appl. Microbiol. Biotechnol. 14: 136
24. Bisping B, Rehm HJ (1984) in Third European Congress on Biotechnology, vol 11, München, FRG
25. Bisping B, Rehm HJ (1986) Appl. Microbiol. and Biotechnol. 23: 174
26. Neish AC, Blackwood AC, Ledingham GA (1945) Canad. J. Research 238: 290
27. Neish AC, Blackwood AC, Ledingham GA (1945) Science 101: 245
28. Blackwood AC, Neish AC, Brown WE, Ledingham GA (1947) Canad. J. Research 258: 56
29. Ledingham GA, Neish AC (1954): In: Underkofler LA, Hickey RJ (eds) Industrial fermentations, vol 2, Chemical Publishing Company, p 27
30. Neish AC, Blackwood AC, Ledingham GA (1947) US Patent 2,432,032
31. Nickerson WJ, Carroll WR (1945), Archi. Biochem. 7: 257
32. Van der Wall JP, Lodder J (eds) (1970) The easts: A taxonomic study, North Holland, Amsterdam, p 555
33. Neish AC (1954) Analytical methods for bacterial fermentations, 2nd revision, NRC, Canada, Bulletin no. 2952
34. Ashworth MRF (1979) Analytical methods for glycerol, Academic, New York
35. Spencer JFT, Sallanas HR (1956) Canad. J. Microbiol. 2: 72
36. Spencer JFT, Shu P (1957) Canad. J. Microbiol. 3: 559
37. Peterson WH, Hendershot WF, Hajny GJ (1985) Appl. Microbiol. 6: 349

38. Hajny GJ, Hendershot WF, Peterson WH (1960) Appl. Microbiol. 8: 5
39. Button DK, Graver JC, Hajny GJ (1966) Appl. Microbiol. 14: 292
40. Onishi H (1961) US Patent 2,986,495
41. Onishi HT (1961) Agricul. Biol. Chemist. 25: 124
42. Onishi H, Saito T (1961) Agricul. Biol. Chemist. 25: 768
43. Onishi H, Saito T (1962) Agricul. Biol. Chemist. 26: 804
44. Onishi H (1963) In: Mrak EM and Steward GF (eds) Advances in food research, vol 12 Academic, New York p 53
45. Vijaikishore P, Karanth NG (1984) Appl. Biochemist. Biotechnol. 9: 243
46. Spencer JFT (1968) In: Hockenhall DJD (ed) Progress in industrial microbiology, vol 7, Publisher Location p 1
47. Spencer JFT, Spencer DM (1978) In: Rose AH (ed) (1978) Economic microbiology, vol 2, Academic, London, p 393
48. Spencer JF, Roxburgh JM, Sallanas HR (1957) US Patent 2,793,981
49. Jain PK, Agarwal GP (1984) In Proceedings of Seventh International Biotechnology Symposium, New Delhi, India, p 414
50. Parekh SR, Pandey NK (1985) Biotech. Bioeng. 27: 1089
51. Agarwal GP, Chattopadhyay A (1986) In: Proceedings of Seventh Australian Biotechnology Symposium, Melbourne, Australia, p 406
52. Spencer JFT, Roxburgh JM, Sallanas HR (1957) J. Agricul. Food Chemist. 5: 64
53. Vijaikishore P, Karanth NG (1986) Process Biochemistry 21: 160
54. Agarwal GP (1985) In: Proceeding of "Biotechnology Asia 85", On Line Publications, Singapore, p 405
55. Sahoo DK, Agarwal GP (1990) Indian Chemical Engineer (in press)
56. Chen BJ, Chi CH (1981) Biotech. Bioeng. 23: 1267
57. Wagmann K (1971) Biochemist. Biophys. Acta 234: 317
58. Ben-Amotz A, Avron M (1973) Plant Physiology 51: 875
59. Ben-Amotz A, Grunwald T (1981) Plant Physiology 67: 613
60. Ben-Amotz A, Avron M (1980) In: Rain DW (ed) (1980) Genetic eng. of osmoregulation: Impact on plant productivity for food, chemicals and energy, Plenum, New York p 91
61. Ben-Amotz A, Sussman I, Avron M (1982) Experientia 38: 49
62. Avron M, Ben-Amotz A (1978) US Patent 4,115,949
63. Muralidhar S (1983) B. Tech. Project Report, IIT, Delhi, India
64. Atkinson B, Mavituna F (eds) (1983) Biochemical engineering and biotechnology handbook, McMillan, London, p 889
65. Mix TJ, Sweck JS, Weinberg M (1978) CEP 74(4): 49
66. Asher DR, Simpson DW (1956) J. Phys. Chemist. 60: 518
67. Prielipp GE, Keller HW (1956) J. American Oil Chemist's Society 33: 103
68. Newman AA (1968), Glycerol, CRC Press, Cleveland
69. Roxburgh JM, Spencer JFT, Sallanas HR (1956) Canad. J. Technol. 34: 248
70. Weiner J (1985) Bio/Technology 3: 41
71. Borowitzka LJ (1986) In: Proceedings of Seventh Australian Biotechnology Symposium, Melbourne, Australia, p 146
72. Murakaml H, Igarashi N (1981) Indus. Eng. Chemist.: Product Res. Dev. 20: 510
73. Desai SV et al. (1972) AIChE Symposium Series 68: 379
74. Vijaikishore P, Karanth NG (1986) Process Biochemist. 21: 54
75. Sols A, Gancedo C, DeLa Fuente G (1971) In: Rose AH, Harrison JS (eds) The yeasts, vol. 2, Academic, London p 271
76. Wilhelmsen JB (1969) Enzymologia 23: 259
77. Weimberg R (1962) Biochem. Biophys. Res. Comm. 8: 442
78. Weimberg R, Orton LL (1963) J. Bacteriol. 86: 805
79. Ingram JM and Wood WA (1965) J. Bacteriol. 89: 1186
80. Spencer JFT, Neish AC, Blackwood AC, Sallanas HR (1956) Canad. J. Biochem. Physiol. 34: 495
81. Brown AD (1979) J. Bacteriol. 118: 769
82. Stanier RY, Doudoroff M, Adelberg EA (1970) The microbial world, 3rd edn, Prentice-Hall, Englewood Cliffs, NJ, p 182

83. Ben-Amotz A, Avron M (1978) In: Caplan SR, Ginzburg M (eds) Energetics and structure of halophilic microorganisms, Elsevier, Amsterdam, p 526
84. Borowitzka LJ, Kessly DS, Brown AD (1977) Arch. Microbiol. 113: 131
85. Borowitzka LJ, Brown AD (1974) Arch. Microbiol. 96: 37
86. Shelef G, Soeder CJ (eds) (1980) Algae biomass production and its use, Elsevier/North-Holland Biomedical, New York
87. Seshadri CV, Thomas S, Bai NJ (ed) (1980) Proceedings of National Workshop on Algal Systems, Indian Society of Biotechnology, BERC, IIT, Delhi, India
88. Aiba S, Humphrey AE, Millis NF (1973) Biochemical engineering, 2nd edn, Academic, New York, p 204
89. Vijaykishore P, Karanth NG (1987) Biotech. Bioeng. 30: 325
90. Wang DIC (1985) In: Ghose TK (ed), Proceedings of Seventh, International Biotechnology Symposium, Delhi, India, p 515
91. Ghose TK, Ghosh B (1985) In: Ghose TK (ed) Proceedings of Seventh International Biotechnology Symposium, Delhi, India p 277

Author Index Volumes 1–41

Acosta Jr., D. see Smith, R. V. Vol. 5, p. 69

Acton, R. T., Lynn, J. D.: Description and Operation of a Large-Scale Mammalian Cell, Suspension Culture Facility. Vol. 7, p. 85

Agarwal, G. P.: Glycerol. Vol. 41, p. 77

Agrawal, P., Lim, H. C.: Analysis of Various Control Schemes for Continuous Bioreactors. Vol. 30, p. 61

Aiba, S.: Growth Kinetics of Photosynthetics Microorganisms. Vol. 23, p. 85

Aiba, S., Nagatani, M.: Separation of Cells from Culture Media. Vol. 1, p. 31

Aiba, S. see Sudo, R. Vol. 29, p. 117

Aiba S., Okabe, M.: A Complementary Approach to Scale-Up. Vol. 7, p. 111

Alfermann, A. W. see Reinhard, E. Vol. 16, p. 49

Anderson, L. A., Phillipson, J. D., Roberts, M. F.: Biosynthesis of Secondary Products by Cell Cultures of Higher Plants. Vol. 31, p. 1

Arnaud, A. see Jallageas, J.-C. Vol. 14, p. 1

Arora, H. L., see Carioca, J. O. B. Vol. 20, p. 153

Asher, Z. see Kosaric, N. Vol. 32, p. 25

Atkinson, B., Daoud, I. S.: Microbial Flocs and Flocculation. Vol. 4, p. 41

Atkinson, B., Fowler, H. W.: The Significance of Microbial Film in Fermenters. Vol. 3, p. 221

Bagnarelli, P., Clementi, M.: Serum-Free Growth of Human Hepatoma Cells. Vol. 34, p. 85

Barford, J.-P., see Harbour, C. Vol. 37, p. 1

Barker, A. A., Somers, P. J.: Biotechnology of Immobilized Multienzyme Systems. Vol. 10, p. 27

Beardmore, D. H. see Fan, L. T. Vol. 14, p. 101

Bedetti, C., Cantafora, A.: Extraction and Purification of Arachidonic Acid Metabolites from Cell Cultures. Vol. 35, p. 47

Belfort, G. see Heath, C. Vol. 34, p. 1

Beker, M. J., Rapoport, A. J.: Conservation of Yeasts by Dehydration. Vol. 35, p. 127

Bell, D. J., Hoare, M., Dunnill, P.: The Formation of Protein Precipitates and their Centrifugal Recovery. Vol. 26, p. 1

Berlin, J., Sasse, F.: Selection and Screening Techniques for Plant Cell Cultures. Vol. 31, p. 99

Binder, H. see Wiesmann, U. Vol. 24, p. 119

Bjare, M.: Serum-Free Cultivation of Lymphoid Cells. Vol. 34, p. 95

Blanch, H. W., Dunn, I. J.: Modelling and Simulation in Biotechnical Engineering. Vol. 3, p. 127

Blanch, H. W., see Moo-Young, M. Vol. 19, p. 1

Blanch, H. W., see Maiorella, B. Vol. 20, p. 43

Blenke, H. see Seipenbusch, R. Vol. 15, p. 1

Blenke, H.: Loop Reactors. Vol. 13, p. 121

Blumauerová, M. see Hostalek, Z. Vol. 3, p. 13

Böhme, P. see Kopperschläger, G. Vol. 25, p. 101

de Boer, L., Dijkhuizen, L.: Microbial and Enzymatic Processes for L-Phenylalanine Production. Vol. 41, p. 1

Bottino, P. J. see Gamborg, O. L. Vol. 19, p. 239

Bowers, L. D., Carr, P. W.: Immobilized Enzymes in Analytical Chemistry. Vol. 15, p. 89

Brandl, H., Gross, R. A., Lenz, R. W., Fuller, R. C.: Plastics from Bacteria and for Bacteria: Poly(β-Hydroxyalkanoates) as Natural, Biocompatible, and Biodegradable Polyesters. Vol. 41, p. 77

Brauer, H.: Power Consumption in Aerated Stirred Tank Reactor Systems. Vol. 13, p. 87

Brodelius, P.: Industrial Applications of Immobilized Biocatalysts. Vol. 10, p. 75

Brosseau, J. D. see Zajic, J. E. Vol. 9, p. 57

Bryant, J.: The Characterization of Mixing in Fermenters. Vol. 5, p. 101

Buchholz, K.: Reaction Engineering Parameters for Immobilized Biocatalysts. Vol. 24, p. 39

Brückmann, A. F., Carrea, G.: Synthesis and Application of Water-Soluble Macromolecular Derivatives of the Redox Coenzymes NAD (H), NADP (H) and FAD. Vol. 39, p. 97

Bungay, H. R.: Biochemical Engineering for Fuel Production in United States. Vol. 20, p. 1

Butler, M.: Growth Limitations in Microcarriers Cultures. Vol. 34, p. 57

Cantafora, A. see Bedetti, C. Vol. 35, p. 47

Carrea, G. see Bückmann, A. F. Vol. 39, p. 97

Chan, Y. K. see Schneider, H. Vol. 27, p. 57

Carioca, J. O. B., Arora, H. L., Khan, A. S.: Biomass Conversion Program in Brazil. Vol. 20, p. 153

Carr, P. W. see Bowers, L. D. Vol. 15, p. 89

Chang, M. M., Chou, T. Y. C., Tsao, G. T.: Structure, Pretreatment, and Hydrolysis of Cellulose. Vol. 20, p. 15

Charles, M.: Technical Aspects of the Rheological Properties of Microbial Cultures. Vol. 8, p. 1

Chen, L. F., see Gong, Ch.-S. Vol. 20, p. 93

Chou, T. Y. C., see Chang, M. M. Vol. 20, p. 15

Cibo-Geigy/Lepetit: Seminar on Topics of Fermentation Microbiology. Vol. 3, p. 1

Claus, R. see Haferburg, D. Vol. 33, p. 53

Clementi, P. see Bagnarelli, P. Vol. 34, p. 85

Cogoli, A., Tschopp, A.: Biotechnology in Space Laboratories. Vol. 22, p. 1

Cooney, C. L. see Koplove, H. M. Vol. 12, p. 1

Costentino, G. P. see Kosaric, N. Vol. 32, p. 1

Daoud, I. S. see Atkinson, B. Vol. 4, p. 41

Das, K. see Ghose, T. K. Vol. 1, p. 55

Davis, P. J. see Smith, R. V. Vol. 14, p. 61

Deckwer, W.-D. see Schumpe, A. Vol. 24, p. 1

Demain, A. L.: Overproduction of Microbial Metabolites and Enzymes due to Alteration of Regulation. Vol. 1, p. 113

Doelle, H. W., Ewings, K. N., Hollywood, N. W.: Regulation of Glucose Metabolism in Bacterial Systems. Vol. 23, p. 1

Dijkhuizen, L. see de Boer, L. Vol. 41, p. 1

Dubois, D. see Miller, O. A. Vol. 39, p. 73

Dunn, I. J. see Blanch, H. W. Vol. 3, p. 127

Dunnill, P. see Bell, D. J. Vol. 26, p. 1

Duvnjak, Z., see Kosaric, N. Vol. 20, p. 119

Duvnjak, Z. see Kosaric, N. Vol. 32, p. 1

Eckenfelder Jr., W. W., Goodman, B. L., Englande, A. J.: Scale-Up of Biological Wastewater Treatment Reactors. Vol. 2, p. 145

Einsele, A., Fiechter, A.: Liquid and Solid Hydrocarbons. Vol. 1, p. 169

Electricwala, A. see Griffiths, J. B. Vol. 34, p. 147

Enari, T. M., Markkanen, P.: Production of Cellulolytic Enzymes by Fungi. Vol. 5, p. 1

Enatsu, T., Shinmyo, A.: In Vitro Synthesis of Enzymes. Physiological Aspects of Microbial Enzyme Production Vol. 9, p. 111

Engels, J., Uhlmann, E.: Gene Synthesis. Vol. 37, p. 73

Englande, A. J. see Eckenfelder Jr., W. W. Vol. 2, p. 145

Eriksson, K. E.: Swedish Developments in Biotechnology Based on Lignocellulose Materials. Vol. 20, p. 193

Esser, K.: Some Aspects of Basic Genetic Research on Fungi and Their Practical Implications. Vol. 3, p. 69

Esser, K., Lang-Hinrichs, Ch.: Molecular Cloning in Heterologous Systems, Vol. 26, p. 143

Ewings, K. N. see Doelle, H. W. Vol. 23, p. 1

Faith, W. T., Neubeck, C. E., Reese, E. T.: Production and Application of Enzymes. Vol. 1, p. 77

Fan, L. S. see Lee, Y. H. Vol. 17, p. 131

Fan, L. T., Lee, Y.-H., Beardmore, D. H.: Major Chemical and Physical Features of Cellulosic Materials as Substrates for Enzymatic Hydrolysis. Vol. 14, p. 101

Fan, L. T., Lee, Y.-H., Gharppuray, M. M.: The Nature of Lignocellulosics and Their Pretreatments for Enzymatic Hydrolysis. Vol. 23, p. 155

Fan, L. T. see Lee, Y.-H. Vol. 17, p. 101 and p. 131

Faust, U., Sittig, W.: Methanol as Carbon Source for Biomass Production in a Loop Reactor. Vol. 17, p. 63

Fiechter, A.: Physical and Chemical Parameters of Microbial Growth. Vol. 30, p. 7

Fiechter, A., Gmünder, F. K.: Metabolic Control of Glucose Degradation in Yeast and Tumor Cells. Vol. 39, p. 1

Fiechter, A. see Einsele, A. Vol. 1, p. 169

Fiechter, A. see Janshekar, H. Vol. 27, p. 119

Finocchiaro, T., Olson, N. F., Richardson, T.: Use of Immobilized Lactase in Milk Systems. Vol. 15, p. 71

Flaschel, E. see Wandrey, C. Vol. 12, p. 147

Flaschel, E., Wandrey, Ch., Kula, M.-R.: Ultrafiltration for the Separation of Biocatalysts. Vol. 26, p. 73

Flickinger, M. C. see Gong, Ch.-S. Vol. 20, p. 93

Fowler, H. W. see Atkinson, B. Vol. 3, p. 221

Fukui, S., Tanaka, A.: Application of Biocatalysts Immobilized by Prepolymer Methods. Vol. 29, p. 1

Fukui, S., Tanaka, A.: Metabolism of Alkanes by Yeasts. Vol. 19, p. 217

Fukui, S., Tanaka, A.: Production of Useful Compounds from Alkane Media in Japan, Vol. 17, p. 1

Fuller, R. C. see Brandl. H. Vol. 41, p. 77

Galzy, P. see Jallageas, J.-C. Vol. 14, p. 1

Gamborg, O. L., Bottino, P. J.: Protoplasts in Genetic Modifications of Plants. Vol. 19, p. 239

Gaudy Jr., A. F., Gaudy, E. T.: Mixed Microbial Populations. Vol. 2, p. 97

Gaudy, E. T. see Gaudy Jr., A. F. Vol. 2, p. 97

Gharpuray, M. M. see Fan, L. T. Vol. 23, p. 155

Ghose, T. K., Das, K.: A Simplified Kinetic Approach to Cellulose-Cellulase System. Vol. 1, p. 55

Ghose, T. K.: Cellulase Biosynthesis and Hydrolysis of Cellulosic Substances. Vol. 6, p. 39

Gmünder, F. K. see Fiechter, A. Vol. 39, p. 1

Gogotov, I. N. see Kondratieva, E. N. Vol. 28, p. 139

Gomez, R. F.: Nucleic Acid Damage in Thermal Inactivation of Vegetative Microorganisms. Vol. 5, p. 49

Gong, Ch.-S. see McCracken, L. D. Vol. 27, p. 33

Gong, Ch.-S., Chen, L. F., Tsao, G. T., Flickinger, M. G.: Conversion of Hemicellulose Carbohydrates, Vol. 20, p. 93

Goodman, B. L. see Eckenfelder Jr., W. W. Vol. 2, p. 145

Graves, D. J., Wu, Y.-T.: The Rational Design of Affinity Chromatography Separation Processes. Vol. 12, p. 219

Griffiths, J. B., Electricwala, A.: Production of Tissue Plasminogen Activators from Animal Cells. Vol. 34, p. 147

Gross, R. A. see Brandl, H. Vol. 41, p. 77

Gutschick, V. P.: Energetics of Microbial Fixation of Dinitrogen. Vol. 21, p. 109

Haferberg, D., Hommel, R., Claus, R., Kleber, H.-P.: Extracellular Microbial Lipids as Biosurfactants. Vol. 33, p. 53

Hahlbrock, K., Schröder, J., Vieregge, J.: Enzyme Regulation in Parsley and Soybean Cell Cultures, Vol. 18, p. 39

Haltmeier, Th.: Biomass Utilization in Switzerland. Vol. 20, p. 189

Hampel, W.: Application of Microcomputers in the Study of Microbial Processes. Vol. 13, p. 1

Harbour, C., Barford, J.-P., Low, K.-S.: Process Development for Hybridoma Cells. Vol. 37, p. 1

Harder, A., Roels, J. A.: Application of Simple Structures Models in Bioengineering. Vol. 21, p. 55

Hardman, N.: Recent Developments in Enzyme and Microbial Biotechnology — Strategies in Bioprocess Design. Vol. 40, p. 1

Harrison, D. E. F., Topiwala, H. H.: Transient and Oscillatory States of Continuous Culture. Vol. 3, p. 167

Heath, C., *Belfort, C.*: Immobilization of Suspended Mammalian Cells: Analysis of Hollow Fiber and Microcapsule Bioreactors. Vol. 34, p. 1

Hedman, P. see Janson, J.-C. Vol. 25, p. 43

Heinzle, E.: Mass Spectrometry for On-line Monitoring of Biotechnological Processes. Vol. 35, p. 1

Ho, Ch., Smith, M. D., Shanahan, J. F.: Carbon Dioxide Transfer in Biochemical Reactors. Vol. 35, p. 83

Hoare, M. see Bell, D. J. Vol. 26, p. 1

Höke, H. see Syldatk, Ch. Vol. 41, p. 29

Hofmann, E. see Kopperschläger, G. Vol. 25, p. 101

Halló, J. see Nyeste, L. Vol. 26, p. 175

Hollywood, N. W. see Doelle, H. W. Vol. 23, p. 1

Wiesmann, U., Binder, H.: Biomass Separation from Liquids by Sedimentation and Centrifugation. Vol. 24, p. 119

Withers, L. A.: Low Temperature Storage of Plant Tissue Cultures. Vol. 18, p. 101

Wu, Y.-T. see Graves, D. J. Vol. 12, p. 219

Yamada, Y.: Photosynthetic Potential of Plant Cell Cultures. Vol. 31, p. 89

Yamanè, T., Shimizu, S.: Fed-batch Techniques in Microbial Processes. Vol. 30, p. 147

Yarovenko, V. L.: Theory and Practice of Continuous Cultivation of Microorganisms in Industrial Alcoholic Processes. Vol. 9, p. 1

Yonaha, K., Soda, K.: Applications of Stereoselectivity of Enzymes: Synthesis of Optically Active Amino Acids and α-Hydroxy Acids, and Stereospecific Isotope-Labeling of Amino Acids, Amines and Coenzymes. Vol. 33, p. 95

Zajic, J. E. see Kosaric, N. Vol. 3, p. 89

Zajic, J. E, see Jack, T. R. Vol. 5, p. 125

Zajic, J. E., Kosaric, N., Brosseau, J. D.: Microbial Production of Hydrogen. Vol. 9, p. 57

Zajic, J. E., Inculet, I. I., Martin, P.: Basic Concepts in Microbial Aerosols. Vol. 22, p. 51

Zlokarnik, M.: Sorption Characteristics for Gas-Liquid Contacting in Mixing Vessels. Vol. 8, p. 133

Zlokarnik, M.: Scale-Up of Surface Aerators for Waste Water Treatment. Vol. 11, p. 157